Trafalgar Square and the Narration of Britishness, 1900–2012

BRITISH IDENTITIES SINCE 1707

Vol. 5

Series Editors:

Professor Paul Ward
School of Music, Humanities and Media,
University of Huddersfield

Professor Richard Finlay
Department of History, University of Strathclyde

PETER LANG

Oxford · Bern · Berlin · Bruxelles · Frankfurt am Main · New York · Wien

Shanti Sumartojo

Trafalgar Square and the Narration of Britishness, 1900–2012

Imagining the Nation

PETER LANG

Oxford · Bern · Berlin · Bruxelles · Frankfurt am Main · New York · Wien

Bibliographic information published by Die Deutsche Nationalbibliothek
Die Deutsche Nationalbibliothek lists this publication in the Deutsche Nationalbibliografie;
detailed bibliographic data is available on the Internet at http://dnb.d-nb.de.

A catalogue record for this book is available from the British Library.

Library of Congress Control Number: 2013944464

ISSN 1664-0284
ISBN 978-3-0343-0814-4

© Peter Lang AG, International Academic Publishers, Bern 2013
Hochfeldstrasse 32, CH-3012 Bern, Switzerland
info@peterlang.com, www.peterlang.com, www.peterlang.net

Printed in Germany

For Adiyono and Riyana

Contents

viii

Acknowledgements

This book began as a PhD thesis at the Australian National University, and so my first thanks must go to my supervisor and colleague Alastair Greig, who has been a consistent supporter and constructive critic and who I am now lucky to call a friend. Also at the ANU, David Bissell, Emmeline Taylor, Kate Lee-Koo and PhD comrades during my candidature have been supportive and generous. John Hutchinson and Greg Noble provided very helpful comments on the thesis that I hope have improved the final product. At Peter Lang, Paul Ward has been invaluable as a series editor, a genuine pleasure to work with who has made this a much better book (although the inevitable mistakes I can only pin on myself). Lucy Melville has been patient and extremely helpful, making the whole process run as smoothly as my miscalculations would allow. I also must thank the following archives and libraries for vital assistance: the Campaign for Nuclear Disarmament collection at the London School of Economics Archives; the Anti-Apartheid Movement collection at Rhodes House in the Bodleian Library, University of Oxford; the Women's Library at London Metropolitan University; the Trustees of the Mass Observation Archive, University of Sussex; the Newspaper Archives of the British Library; the National Archives in the UK; the National Library of Australia; and the Library of the ANU.

If it takes a village to raise a child, it might also take one to write a book. Very special thanks are due to my parents Jojok and Esther Sumartojo, who have been unstinting throughout, who helped immeasurably in the final push to complete the book, and spurred my intellectual curiosity in the very beginning. It is a real pleasure to thank the friends who have provided inspiration and comfort: in England, Julie Lovell, Ivan Meltcalf, Claire O'Brien, Paul Wreaves and Clare Lane; in Australia, Kate Hancock, Agus Purwanto, Jo Lumb, Di Martin, Helen Keane, Fiona Yule, Heather Govender and Kirsten Merlino. Thanks also to Joyce Wellings for a home in the Home Counties and her unfailing kindness and ferocious support, and Rini Sumartojo, who has been an inspiration from afar, as well as a fellow new colonial. Ben Wellings continues to generate energy, creativity, and deep happiness, and it is because of him, quite literally, that I was able to write this book. It is impossible to articulate the support he has given me, but I thank him nonetheless. Of all these wonderful people, however, I dedicate this work to my generous, clever and beautiful children Adiyono and Riyana, whose picture, taken in Trafalgar Square, I proudly display on my desk. Every day they remind me what 'valuable contribution' means. Good on you, kids.

List of Illustrations

CHAPTER I

Introduction

In August 2007, the Taj Mahal materialised in Trafalgar Square. This iconic Indian building featured as part of the Trafalgar Square festival, a project of the London Mayor, Ken Livingstone. This architectural juxtaposition was part of a festival intended to celebrate the creative relationship that London enjoyed with India, and in addition to the reproduction of the Taj Mahal, the three week festival also featured dance and musical performances, and a giant canvas at the foot of Nelson's column that was designed to 're-imagine London as an Indian city'.[1] This festival took place right under the nose of the statue of Sir Henry Havelock, an imperial hero of Victorian Britain whose muscular Christianity was evident in his relief of besieged British women and children in Lucknow during the Indian Uprising of 1857 and its brutality against local civilians.[2] While the relationship between Havelock's London and India was very different from Livingstone's, in choosing to re-imagine London in this way, London's government drew upon a rich history of contact and interaction with Asia, which remains a vital part of the identity of contemporary Britain.

The arranged marriage of these two structures in Trafalgar Square – the Taj Mahal and Nelson's Column – created a spatial juxtaposition of London and India, but also juxtapositions of imperial past and globalised present, nation and individual, and a representation of history and use of place that points to a wide range of questions that this book explores. At the most general level, it asks what Trafalgar Square can tell us about national identity, and explores how the Square has contributed to the construction, maintenance or transformation of British national identity, as well as the potential for the Square to help various national groups resist or alter dominant narratives of Britishness. It also considers the ways that the use of the Square has helped to reframe the national meanings implied by its built forms. The spatial reality of a reproduction of an Indian building in the centre of London is a good place to start this discussion, because it foregrounds how local geographies help to map national identity in ways that are common in urban spaces in London and other cities throughout the world. If London can be reimagined as an Indian city through the use of Trafalgar Square, how else can it be imagined? And what effect does this have on the nation that it represents?

These are important questions because many of us use our cities in different ways. Urban landscapes, even those with the appearance of permanence, are mutable, and many buildings or neighbourhoods have been completely altered over the twentieth century, roughly the period that this book discusses. In London alone, the suburbs mushroomed in the fifty years before World War Two;[3] some areas were completely rebuilt following the war, and tower blocks for low-income housing that were considered the pinnacle of modernity in the 1960s had begun to be dynamited in the 1980s.[4] The year 2012 saw the completion of a new and controversial skyscraper in the Shard, London, and the European Union's tallest building, which commentators contextualised by pointing out that London's landscape is a dynamic and changing one, always in flux.[5]

This was certainly the case during the creation of Trafalgar Square, which began in 1840. The Victorians changed London's landscape drastically, with major infrastructure, transport and housing projects intended

3   Roy Porter, *London: A Social History* (Cambridge, Mass: Harvard University Press, 2001).
4   Jerry White, *London in the 20th Century* (London: Vintage, 2001).
5   Steve Rose, *The Shard: Renzo Piano's great glass elevator* <http://www.guardian. co.uk/artanddesign/2012/jun/13/shard-renzo-piano/print> accessed 10 July 2012.

to accommodate and support the capital city's growing population. From 1817 to 1823, John Nash's redesign of Regent Street took form, intended both to improve the area and to separate the wealthy homes to the west and north from the poor and working-class people to the east in 'confused Soho'.[6] Trafalgar Square was very much a part of a larger landscape of power, domestic as well as international. Victorian and Edwardian visitors to the Square lived in a period when London was the wealthy, busy and diverse centre of an extensive empire, and this global reach was reflected in many aspects of material life, not least the built environment. In 1900, the approximate starting point of this book, London's landscape was, as Jonathan Schneer has shown, an imperial one:

> The public art and architecture of London together reflected and reinforced an impression, an atmosphere, celebrating British heroism on the battlefield, British sovereignty over foreign lands, British wealth and power, in short, British imperialism.[7]

Even as this history is identifiable in central London's layout, it is not simple or static. Many authors have identified the variety, multiplicity and non-hegemonic aspects of London's landscape, and this book shows how the flux and flexibility of national narratives have been played out in London's central urban spaces. In 2012, for example, evidence of this process appeared in Yinka Shonibare MBE's 2012 work for the Square's Fourth Plinth, *Nelson's Ship in a Bottle*. This work presented a modern and multicultural exploration of the enduring effect of imperialism on London and the UK, drawing on a history of imperial trade and commerce, while gently reminding viewers of the exploitative nature of many imperial exchanges.

Shonibare's interpretation of London's relations with the British Empire presented it as part of a fluid, contingent national history. This points to one of the central conceptual approaches that this book takes to Trafalgar Square: that the relationship between the national past and its present is best understood as how the past is reconstructed for the purposes

6   See Porter, *London*, 126–130.

7   Jonathan Schneer, *London 1900: The Imperial Metropolis* (New Haven: Yale University Press, 1999), 19.

of the present.[8] Shonibare's work was a strong reminder both of London's history and the evidence of this history in its buildings, places and layout. By installing his work in Trafalgar Square, he used the larger built environment surrounding the Fourth Plinth to amplify the meaning of his artwork.

Trafalgar Square's prominence for visitors to London, as well as its proximity to Westminster Palace and the Houses of Parliament, government offices along Whitehall, Buckingham Palace and the shops, clubs and cafes of the West End, make it a valuable subject of study. The Square's location in the centre of London is constitutive of its meaning and its value as a nationally visible site. Not least, its imperial symbolism and the implicit link this provides to the highest levels of official power and control have made it highly attractive to groups wishing to challenge this power. As Dennis argues:

> [...] despite, perhaps because of, these attempts to keep Trafalgar Square under state control, it has always been associated more with acts of popular protest than official ceremonial. The attempt to reserve the square for official and approved occasions could even have been a stimulus to protest. The square became a prize of enormous symbolic value.[9]

What makes the contested space of the Square valuable in terms of investigating national identity, however, are the many different types of uses to which it has been put. Alongside Trafalgar Square's well-known history as a protest site is its role in state rituals, such as royal weddings, coronations or jubilees, its importance at moments of national celebration, such as VE Day or victory in football, and its quotidian uses as a meeting point, transit hub or lunch spot for Londoners and visitors for the length of its history. This multiplicity of uses and meanings for different users, as well as its national iconography and official oversight of events there, make it not only a fitting central place for London, but a site in which the meaning of the nation itself can be explored, celebrated or contested.

8   Jeffrey Olick, *States of Memory* (Durham and London: Duke University Press, 2003), 6.

9   Richard Dennis, *Cities in Modernity: Representations and Productions of Metropolitan Space, 1840–1930* (Cambridge: Cambridge University Press, 2008), 163–164.

Overall, this book argues that Trafalgar Square has acted as a proxy for the nation, providing a site in which groups have sought national visibility through visibility in the space of the Square. The Square has also helped groups to imagine and construct British national identity in a way that draws on national history while still remaining flexible in its interpretation.[10] The chapters below will show how protesters using the Square have repeatedly framed their temporary spatial occupation of the site as a means by which to demand national recognition of their causes, and will touch on categories of gender, race, age and urban and imperial identities along the way. Parallel to its history as a protest site is the site's use for official state or metropolitan purposes. Officials have used the Square, often as part of a larger central London landscape, as shorthand for a national space, in events such as Royal celebrations, VE Day, and the 2005 winning of the Olympic Games for London. This mix of uses – Trafalgar Square's symbolic role as a site of national history, its ongoing use for both the everyday and the spectacular, as well as its location at the centre of a larger 'landscape of power' that takes in government, finance and cultural institutions – has made it a powerful site in which national identity is contested, imagined and reproduced. This does not mean, however, that the political changes demanded by protesters have necessarily occurred. For example, 140,000 protesters against the Industrial Relations Bill in 1971, one of the biggest demonstrations ever held in the Square, did not prevent the Bill's enactment (see Chapter 5). While it may be a platform from which to demand social or political change, the Square does not offer these demands a guarantee of success.

The book takes examples from the Square's history to explore how its use draws together national history, social power and the built environment, in a discursive and flexible process of national identity creation. However, this process is not without its boundaries, and the book will also discuss how the construction of national identity within the Square is linked to events of the past as well as being engaged with the specific symbolic language of the space. This approach is grounded in scholarship on national identity.

10   See Paul Ward, *Britishness since 1870* (New York: Routledge, 2004).

National identity and the uses of history

Anderson's seminal notion of the nation as an 'imagined community' proposed some of the specific ways in which the national imaginary is generated
and promulgated, identifying the systemic processes of 'print-capitalism'
that use strategies such as maps to define territory, cultural institutions
such as museums to make certain aspects of cultures official, and ways
of counting and categorising people through the census to create unitary
national identities.[11] In doing so, his main focus rested on the 'top-down'
processes that built a coherent national narrative out of a set of colonial
bureaucratic and economic structures. Hobsbawm's discussion of the 'invention of tradition' highlights a related process, in which powerful groups
work through national social and political structures to serve their own
interests by developing ideologies and symbols to create unified nations
from diverse language and ethnic groups.[12] They identify the important
role of the numerous public monuments, including war memorials, tower-
ing statues and public buildings that reflect narratives of national power.

    These ways of approaching national identity appear to help explain the
ongoing significance of Trafalgar Square. Its name, its Victorian statuary,
its layout – with fountains designed in part to help control the numbers
of people who could gather there – and its position in a larger central
London landscape of imperial power all point to a monolithic and powerful version of the British nation (and Empire).[13] Furthermore, Trafalgar
Square also hints at the importance of a process of 'imagining', the role of
mass cultures, and some of the mechanisms by which national narratives
are created and reproduced. These approaches, however, concentrate on
structure and emphasise an objective cultural form of national identity,

11   Benedict Anderson, *Imagined Communities* (London: Verso, 1991 [1983]).
12   See Eric Hobsbawm, 'Introduction' in Eric Hobsbawm and Terence Ranger, eds, *The
      Invention of Tradition* (Cambridge: Cambridge University Press, 1983).
13   Felix Driver and David Gilbert 'Heart of Empire? Landscape, space and performance
      in imperial London', *Environment and Planning D: Society and Space* (16) 1998, 11–28.

thus implying that such identities are fixed and agreed by all members of the political community. As Graham Day and Andrew Thompson put it, these theories treat 'the nation as a sociological reality [...] a real and unified group', thus side-stepping the possibility of fluidity, contest or multiplicity.[14]

The analysis in this book is framed by an understanding of national identity as a discursive process encompassing many different modes of belonging. This is based on ways of understanding the nation as a framing discourse;[15] a cultural matrix;[16] a set of conflicting narratives;[17] and 'an everyday plebiscite' in which the nation is constantly renegotiated.[18] For example, Özkırımlı's reaction to the notion of the 'invention of tradition' is to acknowledge the constructed nature of specific cultural attributes, or traditions, of nations, while pointing out that nationalism is constructed in many different ways.[19] He argues that while nations hold real and deeply-felt meaning for people, they are dynamic, changing and self-reinventing, based on 'culture [...] that is not a passive inheritance but an active process of creating meaning, not given but constantly defined and reconstituted'.[20] Bhabha similarly suggests that flexible and multiple narratives comprise national identity, and claims that membership of the nation 'must always itself be a process of hybridity, incorporating new "people" in relation to the body politic [and] generating other sites of meaning'.[21]

Other work has addressed the extent to which national identities can be multiple or flexible by focusing on issues of cultural reproduction,

14 Graham Day and Andrew Thompson, *Theorizing Nationalism* (Basingstoke: Palgrave Macmillan, 2004), 12.
15 Umut Özkırımlı, *Contemporary Debates on Nationalism: A Critical Engagement* (Basingstoke: Palgrave Macmillan, 2005).
16 Tim Edensor, *National Identity, Popular Culture and Everyday Life* (Oxford: Berg, 2002).
17 John Hutchinson, *Nations as Zones of Conflict* (London: Sage, 2005).
18 Ernst Renan, 'Qu'est-ce qu'une nation?' (1882) in John Hutchinson and Anthony Smith, eds, *Nationalism* (Oxford: OUP, 1994), 17.
19 Özkırımlı, *Contemporary Debates*, 170.
20 *Ibid.*
21 Homi Bhabha, 'Introduction' in Homi Bhabha, ed., *Nation and Narration* (London: Routledge, 1990).

discourse and narrative, including non-elite groups' relationship with the nation.[22] While this scholarship still links the production of national symbolism to the processes of modernity (such as industrialisation), its main concern is how and by whom this symbolism is constructed and reproduced. These accounts also recognise the 'fluid and dynamic nature'[23] of national identity, and stress the contests and tensions that define it. Calhoun, for example, turns to Foucault's notion of a 'discursive formation' to help capture nationalism's complexity and dynamism, explaining that nationalism is: 'a way of speaking that shapes our consciousness, but also is problematic enough that it keeps generating more issues and questions, keeps propelling us into further talk, keeps producing debates over how to think about it.'[24]

Implicit in this formulation of national identity is the quotidian nature of the process by which it is reproduced. Like a flag hanging limply outside a government building, symbols of the nation are around us all the time, but can easily pass almost unnoticed in everyday life.[25] Other approaches recognise the quotidian as more active, arguing that popular conceptualisations of the nation are significant in reproducing it:

> [National culture] is constantly in a process of becoming, of emerging out of the dynamism of popular culture and everyday life whereby people make and remake connections between the local and the national, between the national and the global, between the everyday and the extraordinary.[26]

Drawing these two points together, Cohen's notion of 'personal nationalism' argues that any study of national identity must take into account the intentions of the producers of national symbolism and ritual, as well as how their audiences read these rituals.[27] In other words, the actions of individuals must be considered in studies of national identity, not least because

---

22   Anthony Smith, 'The limits of everyday nationhood', *Ethnicities* 8 (2008), 564.

23   Montserrat Guibernau, *The Identity of Nations* (Cambridge: Polity Press, 2007), 11.

24   Craig Calhoun, *Nationalism* (Buckingham: Open University Press, 1997), 3.

25   Michael Billig, *Banal Nationalism* (London: Sage, 1995).

26   Edensor, *National Identity*, vii.

27   Anthony Cohen, 'personal nationalism – a Scottish view of some rites, rights, and wrongs', *american ethnologist* 23/4 (1996), 802–815.

'nations and national identity are used by people to position themselves in relation to others'.[28] This framework, then, is built on the relationship between the institutional and official on the one hand, and the vernacular, personal and quotidian on the other. National identity relies on both perspectives to reproduce itself, in a process that includes contingent histories and memories that undergo constant reframing and reimagining.

Having said this, national identity cannot be anything, at any time. While it may be flexible in the long run, able, for example, to incorporate new immigrant populations, reformulations of gender relationships or major, structural economic changes, the discourse is a bounded one. One of the strongest forces in shaping this discourse lies in national histories, which can be reinterpreted and redeployed, with some aspects emphasised or even forgotten, but which cannot be changed altogether. Scholarship on the contingent quality of national histories resonates with ways of thinking about national identity as discursive, progressive and multiple, while still alluding to the limits of reinterpretation. Nora's notion of *lieux de mémoire* is useful here: broadly defined sites, rituals or artefacts of national memory where the past is explicitly, if selectively, evoked and represented.[29] However, *lieux de mémoire* foreground only some aspects of the past, demonstrating a tension between official history, a 'representation of the past', with vernacular or popular memory, 'a perpetually acting phenomenon, a bond tying us to the eternal present'.[30] Nora's identification of these two aspects speaks to the power of institutions to shape national narratives, as well as the resistance, adherence or even indifference to these narratives by the public. As in the discussion of the quotidian, the figure of the individual is ascribed some autonomy to engage with narratives of national identity, even when much more powerful processes are at play.

The uncertain relationship between the nation and its past means that these narratives can both include and exclude, and that nations must

28   Andrew Thompson, 'Nations, national identities and human agency: putting people back into nations', *The Sociological Review* 49/1 (2001), 20.
29   Pierre Nora, 'Between Memory and History: *Les Lieux de Mémoire*', *Representations* 26 (Spring 1989), 7–24.
30   *Ibid.*

tread carefully through history, 'draw[ing] sustenance from their past, yet to be fully themselves must also put it away from them'.[31] History here is fluid, a process of remembering that reconstructs and reproduces the past in light of the aims of the present,[32] and this process includes sites that are 'spaces explicitly designed to impart certain elements of the past – and, by definition, to forget others'.[33] Geographer Tim Cresswell uses the idea of place memory, or 'the ability of place to make the past come to life in the present and thus contribute to the production and reproduction of social meaning'.[34] It follows that by adopting alternative histories of place, its significance can be changed; in other words, the meaning of place changes with different versions of history.[35]

The example of the Taj Mahal in Trafalgar Square that I began with demonstrates the sometimes difficult relationship between the nation and its history, one in which national groups recognise the flexibility, utility and narrative power of history while still being bound by its constraints. We may be able to reimagine London as an Indian city, but this is only possible in the context of a history of interaction that includes violence and exploitation as well as creative cross-fertilisation. If the past provides models of inspiration and a 'lifeline' for the present, it must be 'malleable as well as generously preserved', providing us with 'a heritage with which we continually interact, and which fuses past with present'.[36] In other words, the relationship between the nation and its past provides the symbolism for national identity, and therefore some boundaries to its discourse, without being rigid in how this history is interpreted.

---

31   David Lowenthal, *The Past is a Foreign Country* (Cambridge: Cambridge University Press, 1985), 72.

32   Maurice Halbwachs (trans. by Lewis A. Coser), *On Collective Memory* (Chicago: University of Chicago Press, 1992).

33   Steven Hoelscher and Derek Alderman, 'Memory and place: geographies of a critical relationship', *Social and Cultural Geography* 5/3 (2004), 350.

34   Tim Cresswell, *Place: A short introduction* (Oxford: Blackwell, 2004), 85–86.

35   Doreen Massey, 'Places and their Pasts', *History Workshop Journal* 39 (1995), 182–192.

36   David Lowenthal, *The Past is a Foreign Country* (Cambridge: Cambridge University Press, 1985), 410–411.

National places

The arguably ambivalent relationship between the past and present, the official and popular, the everyday and the nationally spectacular, helps to make place a useful category of analysis for exploring national identity. This is especially true for those sites rich with symbolism of the national past, yet still subject to contemporary use. The large body of scholarship on the relationships between place, monuments and artworks, memory, history, and national identity explore the potential of such sites thoroughly. These approaches coalesce around a range of issues that overlap somewhat with the literature on national identity discussed above.

The flexibility (or otherwise) of national narratives has been a consistent focus of geographical research examining the negotiation and politics of the form, location and symbolism of public artworks.[37] On a metropolitan scale, Till's research on Berlin[38] and Kincaid's on Dublin[39] provide two examples of city-wide processes of remembering and forgetting the past through the urban built environment. Kincaid explores how sites with

37   See for example Nuala Johnson, 'Sculpting heroic histories: celebrating the centenary of the 1798 rebellion in Ireland', *Transactions of the Institute of British Geographers*, 19/1 (1994), 78–93; Rhys Jones 'Relocating nationalism: on the geographies of reproducing nations', *Transactions of the Institute of British Geographers*, 33 (2008), 319–334; Nuala Johnson, 'Cast in stone: monuments, geography and nationalism', *Environment and Planning D: Society and Space*, 13 (1995), 51–65; Tim Edensor, *National Identity*; Dydia DeLyser, '"Thus I salute the Kentucky Daisey's claim": gender, social memory, and the mythic West at a proposed Oklahoma monument', *Cultural Geographies* 15 (2008), 63–94; Steven Hoelscher and Derek Alderman, 'Memory and place: geographies of a critical relationship', *Social and Cultural Geography* 5/3 (2004), 347–355; Michael Heffernan 'For ever England: the Western Front and the politics of remembrance in Britain', *Cultural Geographies* 2 (1995), 293–323; Nuala Johnson, 'The contours of memory in post-conflict societies: enacting public remembrance of the bomb in Omagh, Northern Ireland', *Cultural Geographies*, published online 10 November 2011.

38   Karen Till, *The New Berlin: Memory, Politics, Place* (Minneapolis: University of Minnesota Press, 2005).

39   Andrew Kincaid, *Postcolonial Dublin: Imperial legacies and the Built Environment* (Minneapolis: University of Minnesota Press, 2006).

national symbolic content become important to contests over national
identity, using examples drawn from Dublin's built environment:

> Any attempt to alter the built environment will also change the way the past and the
> future are perceived. Geography is history. The physical landscape bears the traces of
> the past, and all alterations to the built environment are a direct means of rethinking
> and determining which memories survive and which are thwarted or suppressed.[40]

If memory is written in the urban landscape, then the act of collective
memory is a process, 'not an inert and passive thing, but a field of activity
in which past events are selected, reconstructed, maintained, modified and
endowed with political meaning'.[41] This fluidity opens up the possibility
for competing national narratives and identities, and contests over those
identities. Embedded in place, therefore, are power relationships reflected
in part in the political renderings of place's meaning. As discussed above,
Nora identifies the tension between official history and vernacular memory,
and this is evident in official attempts to control which histories are told,
and which groups are cast as oppositional. In this sense, place, including
its representations of history, is implicitly political. What is more, spatial
power relationships can be destabilising and contested, with political tensions inherent in place emerging when social or political groups seek to
promote their agendas to the public:

> there may not be consensus amongst state and local elite groups as to how and if
> these places should be remade, because 'official' agendas vary. Further, different
> social groups, functioning as distinct 'publics' and counter-publics, may interact
> with officials or choose other actions that influence the remaking of these places
> [...] public memory is an activity or process rather than an object or outcome.[42]

The idea of place as the deeply meaningful location of experience and
activity continues to inform how we understand cities and the practices
and relationships they engender. By loosening the bonds between the

---

40   Kincaid, *Postcolonial Dublin*, 228.
41   Edward Said, 'Invention, Memory, and Place' *Critical Inquiry* 26/2(2000), 185.
42   Forest, Benjamin, Johnson, Juliet and Till, Karen, 'Post-totalitarian national identity:
      public memory in Germany and Russia', *Social and Cultural Geography* 5/3 (2004), 358.

meaning of place and its physical form, narrative and interpretation are just as constitutive of place as the built environment.[43] Here, the figure of the user of place helps to determine its nature by engaging in a progressive process that opens up space by linking it to other times and places. In these accounts, place is never fixed in how people approach or experience it, and it always has a relative, contingent quality. It is always constituted in relation to people, their activities and practices. Here, 'practices, mundane everyday practices [...] shape the conduct of human being towards others and themselves in particular sites'.[44] The main concern of this approach is how – within a changing environment – everyday practices help us to engage with the world of representation in the city around us. Here, 'nonrepresentational' practice is made up of small political acts that resist, reinforce, or seek to redefine urban places through everyday use.

Such quotidian, tactical engagements occur in relation to powerful spatial narratives that different users must negotiate in different ways, not least because certain types of uses (and users) are inevitably privileged in any spatial structuring. Place's function as a vehicle for the expression of and resistance to power rests on its relational social constitution, its 'complex web of relations of domination and subordination, of solidarity and cooperation',[45] or the power relationships that are exercised and negotiated through the use of place. Here, it is not use alone that shapes the meaning of place, but rather use in a discursive, dialectical power relationship with dominant cultural forces of control:

> [...] some form of dominance must always be present for culture and/or cultural landscapes to *exist* [...] while struggle is always present in the landscape, it is ultimately the forces of limitation and control, rather than those of interpretation and resistance, that define what culture or the cultural landscape *is*.[46]

---

43 See David Harvey, 'Contested Cities: Social Process and Spatial Form' (1997) in LeGates, Richard and Stout, Frederic, eds, *The City Reader* (Routledge: New York, 2003) and Doreen Massey, *for space* (London: Sage, 2005).

44 Nigel Thrift quoted in Catherine Nash, 'Performativity in practice: some recent work in cultural geography', *Progress in Human Geography* 24 (2000), 655.

45 Doreen Massey, 'Politics and Space/Time' *New Left Review* 196 (Nov.–Dec. 1992), 81.

46 Mitch Rose, 'Landscape and labyrinths', *Geoforum* 33 (2002), 459.

Indeed, in a call for the re-materialisation of geography, Lees argues for an interpretive, ethnographic approach to urban built environments that treats them as an 'active and engaged process of understanding', encompassing both what such environments represent, and how these representations are used, processed, or understood.[47] Others express this in terms of spatialised practice, arguing that place-based expressions of social power occur by means of the use of those places:

> Place is produced by practice that adheres to (ideological) beliefs about what is the appropriate thing to do. But place reproduces the beliefs that produce it in a way that makes them appear natural, self-evident and commonsense [...] thus places are active forces in the reproduction of norms – in the definition of appropriate practice. Place constitutes our beliefs about what is appropriate as much as it is constituted by them.[48]

One potential conclusion here is that the narratives and ideologies activated within urban built environments by spatialised practice are monolithic or deterministic. This is not what I mean to suggest. Instead, it is the discursive unfolding of the meaning of place – the subtle to-and-fro between users and their material environments – that makes place a fruitful subject of analysis: 'The principle is one of multiplication: materiality is never apprehensible in just one state, nor is it static or inert [...] as such, as variously turbulent, interrogative, and excessive, materiality is perpetually beyond itself'.[49] It is important here to recognise the subtlety of spatial power. While some official ideologies might communicate very explicitly about their local or national communities, others are more nuanced, even seductive.[50] The interplay between the users of place and its materiality, including its national representations and symbolism, occurs in a process that parallels the discursive formation of national identity discussed above. Just as the

---

47   Loretta Lees, 'Rematerialising geography: the "new" urban geography', *Progress in Human Geography* 26/1 (2002), 7.

48   Cresswell, *Place*, 15.

49   Ben Anderson and John Wylie, 'On Geography and Materiality', *Environment and Planning A* 41 (2009), 332.

50   See John Allen, 'Ambient power: Berlin's Potsdamer Platz and the seductive logic of public spaces', *Urban Studies* 43/2 (2006), 441–455.

nation is an ongoing process, so people change place through their use of it from minute to minute, user to user. It is never entirely fixed in how people approach or experience it and place always has a relative quality.

Urban publics

Implicit in the relationship between power and place is the question of what is meant by the term 'public'. As with Nora's expansive definition of what a 'site' of memory might be (including rituals, literary texts, monumental places, and cultural history), Warner similarly links the notion of a 'public' to a particular text, arguing that 'a public is a space of discourse organised by nothing other than discourse itself'.[51] Warner's detailed explanation of what he means by public is based on its discursive, contingent, reflexive and multivalent qualities, placing 'public address' as its core, claiming that a public 'exists by virtue of being addressed'. For others, the text around which a public can coalesce includes particular public spaces and their material environments.[52] Public place is necessary for a flourishing public sphere of democratic practice to exist. In a cosmopolitan city such as London, public sites are 'perceived as political spaces. Serving on the one hand to exercise and claim civic rights, public spaces are on the other regulated in terms of who may appear in, or intervene in the decision-making about, these sites'.[53]

If the use of place is important in determining its meanings, then what types of use are important? Several different perspectives emerge from the literature, highlighting the significance of both spectacular and quieter, everyday ways of using place. Trafalgar Square has hosted a range of public

51 Michael Warner, 'Publics and Counterpublics', *Public Culture* 14/1 (2002), 50.
52 Iveson, Kurt, *Publics and the City* (Oxford: Blackwell: 2007), 12.
53 Uta Staiger, 'Cities, citizenship, contested cultures: Berlin's Palace of the Republic and the politics of the public sphere', *Cultural Geographies* 16 (2009), 312.

events, from celebrations and festivals to rallies and violent protests, and
has been the setting for the unusual, extraordinary and spectacular. In this
sense, it can be understood as a stage on which groups invoke nation and
history and reach a wider public with their messages.[54] This conceptualisa-
tion of place as a stage, dynamic and in-use, highlights the active process
of the reproduction of meaning in place. Hoelscher and Alderman extend
this idea of place as a 'stage' to less spectacular, although not quite banal,
activities:

> Through bodily repetition and the intensification of everyday acts that otherwise
> remain submerged in the mundane order of things, performances like rituals, festivals,
> pageants, public dramas and civic ceremonies serve as a chief way in which societies
> remember [...]. Civic celebrations [...] are always embedded in place and inevitably
> raise important questions about the struggle of various groups to define the centre
> of urban politics and urban life.[55]

Here, certain places act as stages for national political activities that
range from the spectacular and official to the everyday and vernacular.
Furthermore, through a process of 'symbolic accretion', the weight of accu-
mulated use deepens the meaning of national places. According to Dwyer,
'activists use symbolic accretion as a strategy for burnishing the reputation
of their cause via proximity, both actual and metaphorical, within an established memorial landscape [...]. [Monumental places are] political resources,
laden with authorial intentions, textual strategies and readers' interpretations.'[56] However, as discussed above, place is not a static text, but is reproduced and remade through use, as sites are both shaped by the groups that
use them and influence those groups. This scholarship characterises place
as practiced and reproduced in a process that binds contemporary users
to historical narrative, while also being shaped by previous uses of place.

54   Joshua Hagen and Robert Ostergren, 'Spectacle, architecture and place at the
     Nuremberg party rallies: projecting a Nazi vision of past, present and future', *Cultural
     Geographies* 13 (2006), 157–158.
55   Hoelscher and Alderman, 'Memory and place', 350.
56   Owen Dwyer, 'Symbolic accretion and commemoration' *Social and Cultural
     Geography* 5/3 (2004), 420–422.

As I will show below, this has certainly been the case in Trafalgar Square, particularly with the accreted meanings associated with its ongoing use for public protest, often with national aims.

Other perspectives on the use of place raise the notion of 'spectating' as a practice which can occur in many different types of places, and which, despite its apparent passivity, is one way that people can exercise agency. Spectating reproduces meaning, and is 'a complex, ambivalent relation to visual objects [such as the built environment], in which the psychic and the discursive quite often, if not usually, displace questions of composition or veracity [...] there's never such a thing as a gaze in some kind of vacuum'.[57] In other words, spectating or looking is strongly informed by the perspective of the spectator. As with other uses, place is made by how it is regarded or looked at, and the way place is constituted by the spectator is informed by their particular social perspective – gay, black, female, parental or middle-class, for example. This is important for a national site such as Trafalgar Square because it reinforces the flexibility and mutability of both the built environment and the national narratives that it can represent.

As discussed above, one theme in scholarship on place is the importance of power relationships that are exercised and negotiated through its use. This is variously expressed as an uneasy tension between symbolic representation and use of place, between elite control of place and public reactions to that control, even between official and vernacular approaches to history. This tension is implicit in the notion of symbolic accretion. This discursive, dialectical power relationship helps to shape the meaning of place; as an aspect of material culture, the built environment is as meaningful as other forms of culture for the individual, and its use helps to regulate and define society, including national identity. In this book, I will show how this process has unfolded in Trafalgar Square and how Britishness has been constructed, represented and re-imagined in both the built environment and the use of the Square.

---

57   In Peter Merriman *et al*, 'Landscape, mobility, practice' *Social and Cultural Geography* 9/2 (2008), 201.

Plan of the book

This book is organised chronologically, because the order in which things have happened in Trafalgar Square demonstrates my arguments about history, use of place and accreted meaning, and the role of the built environment in constructing national identity. Overall, in each chapter I look closely at an event or series of events to draw out the specific ways that the environment of the Square helped to shape the national narratives at stake. This material will be contextualised by a broader discussion of London and Britain that links the events in the Square to the national social and political contests of the day. This approach means that I do not consider *everything* that occurred in the Square, but have instead tried to select examples that give a sense of the scope and range of national narratives on display there.

Chapter 2 introduces the reader to Trafalgar Square by explaining the conditions of the Square's construction, including some of the controversies that accompanied it, and its early uses. Importantly, this includes a series of labour protests that culminated in 1887 with the death of a protester and controversy about police handling of the event. This chapter also explains the larger 'landscape of power' in which the Square is situated, and draws out some of the social and political tensions that were often on display there. I also discuss the symbolism of the statues and monuments in the Square, building up a picture of the space that underpins the rest of the book.

The early history of conflict in the Square would become a reference point for later protesters, as Chapter 3 shows. Here, I examine the use of the Square by female suffrage protesters, framing their actions with the challenge to the notion of the urban public that such women represented. Groups such as the Women's Social and Political Union used the Square as part of a purposeful campaign to raise the national visibility of women by making themselves visible in the public spaces of urban Britain. As such, the space of the Square acted as a national space, a proxy for the nation. Female suffrage protesters attempted to increase their prominence by activating the Square (and other parts of central London) as explicitly national spaces. Furthermore, the Suffragettes' use of the Square occurred at a time when women were entering the public sphere in many different ways, and

many of their protest displays played upon aspects of Edwardian femininity that emphasised visual beauty and respectability whilst still demanding
national political inclusion.

Chapter 4 concerns the interwar period and the Second World War.
In it, I discuss the 1937 coronation of George VI, drawing in part from
first-hand accounts of the event in the Mass-Observation Archive to understand how the site was used for spectacular occasions of state, as much as
for highly public protests. At the centre of an unabashedly imperial landscape, Trafalgar Square helped to buttress the legitimacy and power of the
monarchy, and, by extension, the official controls of the site. I also discuss
the uses of the Square during the Second World War, drawing out how it
formed part of a landscape reshaped by the Blitz of 1940–1941, as well as
other wartime alternations to urban space through propaganda or safety
campaigns, such as the blackout. I discuss how London's 'landscape of fear'
challenged the conventional notions of stoicism and resilience exemplified
in propaganda films such as *London Can Take It* (1940). The celebrations
of VE Day in 1945 conclude this chapter, and I show how the floodlighting of many central London landmarks at the end of the war was, for many
observers, less about the survival of the nation and more about the affective
thrill of being in a crowd or the spectacle of the dazzling illuminations.

Chapter 5 opens in an exhausted post-war London still subject to austerity measures, and tracks the many protests in the Square that went to the
heart of post-colonial economic and political transition. This was a time in
which the Square demonstrated how national identity is constructed with
an eye to the outside world. I focus chiefly on the Campaign for Nuclear
Disarmament, with its orientation towards Britain's relationship with both
the US and the USSR, and the long-running campaign against apartheid
South Africa that continued for decades, both in the Square and just outside
it, at South Africa House. Throughout this period, I show how the Square
was often a site that was understood as a *national* one. In contrast to, for
example, the streets of Brixton or Notting Hill, where violent conflicts
over racial tensions were often described as local problems, protests in the
Square were constructed as national, lending protesters a wider-ranging
public voice. This chapter concludes in 1990, when the Poll Tax Riot became
the biggest and most violent conflict that the Square had seen in decades.

Finally, Chapter 6 brings us into twenty-first century London. By 2003, the Square had been redesigned, with the road between it and the National Gallery to the north pedestrianised. This was part of a strategy by London Mayor Ken Livingstone to enliven the space and make it into a more accessible gathering place. However, the appearance of greater physical accessibility was accompanied by tighter control over who could use the Square and for what purpose. The introduction of 'Heritage Wardens' and the systematic removal of pigeons from the space arguably made it safer and cleaner, but possibly subject to fewer instances of spontaneous, unprescribed use. Chapter 5 discusses the 2005 celebrations of London's successful bid for the 2012 Olympic Games, followed a week later by the vigil for the victims of the 7 July bombings on the London Underground and bus system. For both these events, London's racial and cultural diversity was an important point of emphasis, and the Square's imperial landscape was reinterpreted as a stage for official display of London's 'super-diversity'. I also show how this environment was further reinterpreted by the collection of contemporary artworks mounted on the Fourth Plinth, the pedestal in the northwest corner of the Square that has never been permanently filled. I discuss several of the works, and their comments on national identity, concluding with Shonibare's treatment of British history, imperialism, and urban diversity in his popular installation, *Nelson's Ship in a Bottle*.

Overall, the book argues that Trafalgar Square has helped to imagine and construct Britishness, providing context for this process through its built environment and its history of use, while at the same time remaining flexible and receptive to the demands of national redefinition. Despite its appearance of imperial inflexibility, Trafalgar Square reveals itself as a highly mutable space that allows protesters, spectators, passers-by and observers the opportunity for national visibility through its use. Thus, Trafalgar Square has allowed many different groups and individuals, for many different purposes, to make a claim on national membership.

CHAPTER 2

Introducing the Square

Trafalgar Square simultaneously attracts a steady flow of tourists, is a site for special events, and is a place for Londoners to meet, pause or pass through on their way to other places. Its national representations are both fixed and fluid: solid in the Craigleith sandstone and bronze of Nelson's Column, but flexible in the use of these elements as backdrops for dance performances, children's play, or protest rallies. For a previous Director of the National Gallery, Charles Smith, the Square's multiple roles are inherent to the space, if slightly regrettable: 'It is perhaps too often spoiled in appearance by temporary festivals and the ephemeral rubbish they generate, but the combination of history, grandeur and public protest is part of the psyche of the Square'.[1] This chapter introduces the Square, starting with the history of its construction and what it meant to visitors, residents and officials in its early years. It discusses the Square's position in the larger metropolitan landscape of power and explains some of the early conflicts over access to it. This diversity of uses supports one of the main arguments that I will make throughout the book: that, for many users, visibility in the Square provided national visibility that allowed different groups to stake a claim in the nation, and to present a range of ways of constructing Britishness. The environment and history of the Square played a role in this process, however, setting the terms under which these claims could be made.

The range of different uses of the Square to some degree represents its place in a larger web of spatial networks or paths within central London, Britain, and the wider Empire. Perhaps most obviously, Trafalgar Square has an inescapably imperial visual language, evident in the nineteenth-century

1   Jean Hood, *Trafalgar Square: A Visual History of London's Landmark Through Time* (London: Batsford, 2005), 7.

military figures depicted there. As such, it can be conceptualised as the heart of a landscape of global power that connects central London to places around the world, 'celebrating British heroism on the battlefield, British sovereignty over foreign lands, British wealth and power, in short, British imperialism'.[2] It is also a long-standing node of social, commercial and physical transit and exchange. Just opposite Trafalgar Square, in the entrance to Whitehall on Charing Cross, a few dozen metres from Nelson's Column, is a bronze equestrian statue of Charles I that has stood there since 1676.[3] Set into its base is a plaque that marks the point from which all distances in London are measured, indicating the site's role as a national spatial milestone that marks the geographical centre of London, and by extension the centre of Britain and of the former Empire.

The site was also home to the nearby Golden Cross Hotel, a major terminus for coaches from across the United Kingdom that was linked to areas outside the metropole; during the 'early decades of the [nineteenth] century, the Golden Cross and the inns nearby were filled with newcomers and visitors just arrived or just about to depart'.[4] For many early visitors to London from the British hinterland, this part of the city was the first they encountered directly, and the many inns near the site provided popular meeting places for new Londoners recently arrived from the provinces to meet others from their local regions. The area also saw public congregation and the expression of political views before the existence of Trafalgar Square, in large part because of its proximity to the Palace of Westminster. Some of these were orderly protests, while others ended in violent confrontation between participants and forces of the state, shaping it as a 'site of a continuing sparring match between the state and the people'.[5] The Square has also been used for less dramatic purposes. Every day, people use it to

2   Jonathan Schneer, *London 1900: The Imperial Metropolis* (New Haven: Yale University Press, 1999), 19.

3   Hood, *Trafalgar Square*, 13–14.

4   Jerry White, *London in the 19th Century: A Human Awful Wonder of God* (London: Vintage, 2007), 101.

5   Rodney Mace, *Trafalgar Square: Emblem of Empire* (London: Lawrence and Wishart, 2005 [1976]), 23.

rest or play, as a tourist destination or transit point, as well as for advertising, community celebrations or meeting friends. Some people stop or move through it without noticing it at all. Geographer Tim Edensor links the ongoing multiple uses of the Square to the expression or experience of national identity:

> Certain spaces of assembly inevitably associated with national identity, such as [...] Trafalgar Square in London [...] are venues for seething motion, a multiplicity of activities, identities and sights. In contrast to the rather purified, single-purpose spaces of state power, they are more inclusive realms which allow for the play of cultural diversity. They provide an unfixed space in which tourists and inhabitants mingle, people picnic and protest, gaze and perform music or magic, sell goods and services, and simply 'hang out'.[6]

This availability of the space to many different uses means its significance can change for different groups and individuals. However, its symbolism and history frame activities there with an official national discourse deeply rooted in imperial wealth and ambition. These ambitions included the desire for a capital city that outshone European rivals.

'Let it be called Trafalgar Square'

The creation of Trafalgar Square had its roots in early nineteenth-century changes to central London that included the development of large areas of the West End, with new streets and vistas opening up the crowded, busy city. Proposed and developed by planner and architect John Nash, with the support of George IV, these plans included the regulation of spatial access of the lower classes to more wealthy areas of West London. In 1812, when Nash proposed his design changes to the streets and buildings around

---

6   Edensor, *National Identity*, 48. Other spaces that Edensor identifies as similarly associated with national identity are Times Square in New York City, Marrakesh's Djma-el-Fna, India Gate in Bombay, and the Zocalo in Mexico City.

Charing Cross, the Trafalgar Square site was at a crossroads of London classes: to the south and west were government buildings and upper class homes, while to the north were some of the city's poorest neighborhoods. Nash's dramatic changes to this part of London were centred on the creation of Regent Street between 1817 and 1823, a fashionable 'parade and shopping centre' with 'palace-like shops'. Upon seeing it, one 1814 visitor simply declared 'What a lot of stuff!'[7] By building along this route, Nash quarantined Soho to the east from the fashionable properties to the west, making 'London's grandest thoroughfare [...] its social barrier, with Portland Place and Regent Street screening the fashionable West End from *déclassé* quarters'.[8] This aspect of the site's location meant that control of access by the less well-off was a consideration for authorities, and demonstrates that from its beginning the site itself was literally contested ground in terms of class, poverty and, eventually, politics.

Throughout the late 1820s, the Office of Woods, Forests, Land Revenues, Works and Public Buildings gradually acquired the properties making up the Square, and in 1830 William IV approved the name 'Trafalgar Square'. The National Gallery was completed in 1838, while the site that would become Trafalgar Square was laid out between 1829 and 1841.[9] By April 1840, architect Charles Barry had been commissioned to construct the Square using plans that were similar to previous ones drawn up by the architect of the National Gallery, William Wilkins.[10]

The proximity of extreme poverty and wealth to the site continued to vex the Square's designers and funders. From its earliest stages it appears that control and use of the Square, especially by protesters of the lower classes, was a consideration and a concern for planners. In an 1841 letter to the Treasury, for example, the First Commissioner of the Office of Public Buildings, who had administrative responsibility for the Square, alluded to official concern about the potential for the Square to be used for popular

7   Roy Porter, *London: A Social History* (Cambridge, Mass: Harvard University Press, 1994), 130.
8   Porter, *London*, 127.
9   White, *London in the 19th Century*, 26.
10   See Mace, *Trafalgar Square*, chapter 1.

protest, referring to 'evils of a generally objectionable character [that] may be anticipated from leaving so large a space in this particular quarter of the metropolis'.[11] Given London's role as a centre for a range of radical movements in the 1830s, including the growing Chartist movement, the possibility of the use of the space by protesters was well-founded.[12] Charles Barry, the Square's designer, agreed and fountains were added to both improve the Square's amenity and decrease the open space, which could potentially attract large numbers of protesters.[13]

When the Square was being designed, London had no major memorial to mark Admiral Nelson's victory in the Battle of Trafalgar on 21 October 1805, and after news of the battle and Nelson's death reached London, fundraising began for a monument to his achievements.[14] However, it was not until [1838–] around the time when the National Gallery was completed and plans for the Square were being commissioned – that the Nelson Memorial Committee, formed to raise money and commission a monument, won the approval of the Chancellor of the Exchequer to use Trafalgar Square for the new memorial.[15] By January 1839, William Railton had won the right to design the Nelson monument in a contentious competition that included letters of complaint from disappointed competitors. The magazine *Art Union* objected to Railton's design of Nelson's attire, claiming that the costume would 'hand down specimens of the bad taste of a nation' and that '[marble] should exhibit the attributes of the mind, not the decoration of the body'.[16] Eventually, however, the Committee finalised its decision, and responsibility for the different sculptural elements – the statue and column, the reliefs at the base of the column depicting Nelson's major achievements, and the lions surrounding the monument – were each assigned to a different artist.

11   In Mace, *Trafalgar Square*, 87.
12   White, *London in the 19th Century*, 364–365.
13   Hood, *Trafalgar Square*, 49.
14   Mace, *Trafalgar Square*, 58.
15   *Ibid.*
16   *Ibid.*

The construction of the Square did not run smoothly, and in July 1840 Parliament set up a Select Committee to investigate the entire project, on the basis of concerns over rising costs and the possibility that the government might have to pay for any funding shortfall.[17] The Committee appeared to be looking for reasons to stop the project, as demonstrated by their views on the potential for Nelson's Column to 'block up' the valuable open space of the Square:

> [...] it is undesirable that the Nelson Column should be placed in the situation which is at present selected. If it is desirable in a great city to suggest the idea of space, and having obtained space, not to block it up again; if the general architectural effect of Trafalgar Square or of the buildings around it, is to be considered [...] the situation at present selected for the Nelson Column is most unfortunate.[18]

However, by 3 November 1843, despite delays and challenges, Nelson's statue was in position on the column. The *London Illustrated News* covered its labourious ascent and placement in detail, alluding to the symbolic value of the monument: 'may the great memorial [...] be as a pharos to the public spirit in all-coming time'.[19] Here, in the figure of Nelson, the monument was introduced as one that would speak to future generations of the imperial dominance of Britain, and use urban space to express this power. However, not everybody was satisfied with the monument. In 1844, Nottinghamshire MP Henry Knight thought the column too short and the statue of Nelson a 'figure of fun' rather than a suitably heroic depiction of the Admiral. In scathing terms, he described it as 'another architectural disgrace to this metropolis [... that] did as much harm as possible to the finest situation in the world; that magnificent [Trafalgar] square'.[20] Knight's tirade goes on to hint at the way public monuments can be subject to a wide range of interpretations:

17   Hood, *Trafalgar Square*, 50 and Mace, *Trafalgar Square*, 69.
18   In Mace, *Trafalgar Square*, 71.
19   *Ibid.*
20   House of Commons Debate, 22 July 1844 vol 76 cc1246–1253.

[...] Frenchmen, as he had been told, when they came to London, mistook the statue for that of Napoleon, and he had been credibly informed that this imaginary generosity on the part of the British Nation has considerably allayed the irritation against this country which had recently prevailed in France.[21]

Although the Nelson Memorial Committee disbanded in 1844, it took until July 1858 for the Office of Woods, Forests and Land Revenues, which was now responsible for the Square, to ask Sir Edwin Landseer to sculpt and cast the four lions at the base of the column that were the final element of Railton's design. Landseer was a painter who had never sculpted before; although it took ten years to complete the commission, and the lions were quietly installed in 1867, the last element of Trafalgar Square to be completed.[22] The whole process of building the site had taken almost thirty years, and was accompanied by conflict over the choice of designs and artists, over-expenditure and poor workmanship. By 1870, the Square was reckoned to have fallen victim to a massive cost overrun: at least £50,000 had been spent on the monument, compared with the £20,000–£30,000 that the original 1838–1839 competition had specified.[23] According to Hood, 'Nelson would have been astonished at the controversy, inefficiency, fraud and incompetence that had delayed the monument and [increased] the costs'.[24]

The history of the Square's design and construction shows how it was subject to competing concerns, with the final outcome a product of many different relationships and negotiations. Its construction was perceived as inefficient and open to corruption; the lions were slow to arrive; the fountains had to be redesigned and were initially used as rubbish bins by the public; and the column itself was said to ruin the view of the National Gallery. However, the Square was conceptualised as a 'national' place from its beginnings, demonstrated by its dedication to Admiral Nelson and the Battle of Trafalgar. With the choice of name and central monument,

21 *Ibid.*
22 Hood, *Trafalgar Square*, 63.
23 Mace, *Trafalgar Square*, 109.
24 Hood, *Trafalgar Square*, 64.

a martial and masculine national narrative was built into the space. This was overlaid with the realities of the social, class and political tensions of the early nineteenth century. Cost overruns, controversial selection processes for the Nelson Monument's design, and poor quality and delays in some elements were banal but important factors in shaping the Square's finished form.

Statues in the Square

When the Square and Nelson's Column were finally completed, controversy accompanied the creation and placement of new statues and monuments within it. This included disagreement over who should be depicted in the Square and where proposed statues should be placed. Some statues were created specifically for the site, whereas others were brought from elsewhere, and some have been removed since their initial placement. When considering plans for the monument, the Nelson Memorial Committee took the view that 'Poets, artists and politicians had their place in Westminster Abbey; Trafalgar Square should be for the men who took up arms for Queen and country'.[25] The first representations of these figures appeared between the 1840s and the 1880s, when the statues of George IV (1843), Charles Napier (1856) and Henry Havelock (1861) were installed, and in 1888 a statue of Charles George Gordon. Best known for his 'martyrdom' in Khartoum, Sudan in 1885, Gordon's statue stood between the Square's fountains, but was moved during World War Two to make way for the display of a Lancaster bomber.[26]

The statues of Napier and Havelock underscore the imperial flavour of the Square, and each statue references its subject's activities in the Indian subcontinent.[27] Major-General Charles James Napier, for example, became the Governor of the province of Sind in 1843, created a police

---

25   *Ibid.*
26   Mace, *Trafalgar Square*, 125–126.
27   Deborah Cherry, 'Statues in the Square: Hauntings at the Heart of Empire', *Art History* 29/4 (2006), 678.

force, reformed the local civil service and prided himself on delivering justice for the poor and for women. He was a career soldier who believed in the inevitability of imperial rule in the Indian sub-continent on the basis of Britain's superior civilisation, but was also harshly criticised for lack of knowledge of the area and disrespect for local customs.[28] Although not originally intended for the site, his statue was installed in Trafalgar Square in 1856, three years after his death.

In contrast, Henry Havelock's statue was specifically designed for the Square. Havelock captured the public imagination when he relieved the town of Lucknow, which had been besieged during the Indian Uprising (or Mutiny, as it was known to the British at the time) of 1857. The Uprising resulted in the murder of British civilian women and children by mutinous troops in Cawnpore (now Kanpur), an incident which riveted the Victorian public with its lurid accounts (mostly fictional or exaggerated) of sexual violence that 'upset habitual hierarchies of British rule'.[29] Despite his violent treatment of civilians in the countryside surrounding the besieged Lucknow, Havelock was perceived by the public at home as the antithesis of this barbarity, an archetypical muscular Christian.[30] His religious devotion was publicly well-known, and after his death from dysentery in November 1857, he was grieved by the Victorian public as a Christian hero. According to *The Times*, his death was 'a national misfortune. It has fallen upon the British public with the suddenness of a thunderclap [...] and the regret expressed by all, both high and low, is such as can scarcely be surpassed by the lamentation of the nation on learning the death of Nelson in the honour of victory [...].'[31]

National misfortune aside, a more prosaic and practical aspect of the Square hints at the maps that helped British heroes such as Havelock measure, value and thereby control their Empire: a set of imperial standards of length was installed in the Square in 1876. If Charing Cross is the

28 Cherry, *Statues*, 678 and Mace, *Trafalgar Square*, 113.
29 Gautam Chakravarty, *The Indian Mutiny and the British Imagination* (Cambridge: Cambridge University Press, 2006), 38.
30 Schama, *A History*, 252–255.
31 'General Havelock', *The Times* (8 January 1858), 7.

point from which all distances in London, and by extension Britain and the Empire, were measured, these are the standard lengths used to measure them. In addition to the statues, the plaque connects present-day users of the Square to a narrative of the national past that emphasises the martial and imperial.

According to Driver and Gilbert, imperial symbolism is particularly evident in the central London area around Trafalgar Square: 'This triangular area – with Buckingham Palace, Trafalgar Square, and the Houses of Parliament at its three corners and the ceremonial routes of the Mall and Whitehall along two sides – is probably the most celebrated of all sites of 'imperial' London'.[32] Before the Second World War, for example, tourist guidebooks to London emphasised London's role as the centre of the world's largest empire. In one, 'Trafalgar Square was "truly the centre of Empire" not only because of its historical significance, but also because Canada and South Africa had "chosen wisely to place their London home here"'.[33] A 1932 poster for the London Underground called on passengers to 'Visit the Empire – by London's Underground', linking individual stops to imperial sites: 'Aldwych for India House', 'Temple for Australia House', and 'South Kensington for Imperial Institute'. Here, London's geography symbolised the Empire: 'London was not merely the heart of a global empire; it was the place in which an enormous variety of imperial sights could be seen.'[34]

While these examples suggest that Empire was an important part of how London itself was both represented and understood, other accounts suggest that it was not significant for the British public. Bernard Porter has argued that the Empire was not very important to people in Britain 'for most of the time that [Britain] was acquiring and ruling the greatest empire ever', and that its subsequent representations of the late nineteenth and early twentieth centuries have had Empire 'inserted' as culturally significant.[35]

---

32   Felix Driver and David Gilbert, eds, *Imperial Cities: landscape, display and identity* (Manchester: Manchester University Press, 1995), 17.

33   David Gilbert, '"London in all its glory – or how to enjoy London": guidebook representations of imperial London', *Journal of Historical Geography* 25 /3 (1999), 279.

34   Driver and Gilbert, *Imperial Cities*, 1.

35   Bernard Porter, *The Absent-Minded Imperialists: Empire, Society and Culture in Britain* (Oxford: OUP, 2004), 3.

Certainly, before the Second World War, Colonial Office surveys found evidence of widespread ignorance of the Empire, which continues today.[36] Andrew Thompson's nuanced account of the impact of imperialism on Britain after 1850, however, argues that the diversity and plurality of both the Empire and Britain meant that the impact of empire was 'complex and (at times) contradictory' with the range and extent of different types of relationship with the Empire mitigating against 'any single or monolithic "imperial culture" in Britain'.[37] Public reaction notwithstanding, the British Empire was and still is evident in the built environment of central London, and was part of how the Square's designers, builders and sponsors wanted the public to understand the space. As art historian Richard Williams argues, 'What is meant to impress the beholder, to hold him in awe as he gazes on the square, is this history of imperial conquest, a history that is reiterated aesthetically by its grandiose neoclassical architecture'.[38]

Whereas Napier, Havelock and the standards of length provide evidence of the imperial narrative in Trafalgar Square, the busts of World War One naval commanders, Admirals John Rushworth Jellicoe and David Beatty, who both participated in the battle of Jutland on 31 May 1916, emphasise the naval symbolism established by the Nelson monument.[39] These busts were part of a 'replanned' Trafalgar Square that opened in October 1948, additions that were delayed by the Second World War. The fountains were also refurbished, and were floodlit in a display that the Danish royal family visited on a sightseeing tour of the capital.[40] The fountains created enough of a spectacle to merit the following whimsical description:

36 John MacKenzie, ed., *Imperialism and Popular Culture* (Manchester: Manchester University Press, 1986), 7.

37 Andrew Thompson, *The Empire Strikes Back? The Impact of Imperialism on Britain from the Mid-nineteenth Century* (Harlow: Pearson, 2005), 4–5.

38 Richard Williams, *The Anxious City: English urbanism in the late twentieth century* (London: Routledge, 2004), 133.

39 Hood, *Trafalgar Square*, 91.

40 'Royal Sightseers: Drive Last Night Round Trafalgar Square', *The Manchester Guardian* (25 October 1948), 5.

They look exactly like something to eat at some superlative children's party – coloured jellies in ingenious bowls which not only are entrancing to look at but change their colours before you have time to dig a spoon into them.[41]

That the redesigned fountains and the busts of naval commanders should be unveiled at the same time shows how the site could simultaneously be a source of both popular delight and a display of official power. In a counter-balance to the 'coloured jelly' fountains, the continuity between new heroes and old was made explicit in the case of Andrew Browne Cunningham, who served as Admiral-in-Command in the Mediterranean during World War Two. His bust, located in the Square in 1966, sits alongside those of Jellicoe and Beatty.[42] He was described by US General Eisenhower, with whom Cunningham served, as a 'Nelsonian type of admiral [...] he always thought in terms of attack, never of defence'.[43] Along with the statue of Nelson, these figures act as a reminder of the importance of naval power in British history, and spatially link the twentieth century naval commanders to the nineteenth-century one for whom the Square was named. However, as with the construction of other parts of the Square, the presence of these elements also reflects the many changes which have taken place in the site, and represents some of the controversies over those changes. Disagreement and prevarication accompanied the placing of several of the statues, with some, such as George IV and Charles Napier, arriving in the Square after several alternative locations were considered. As described above, Gordon of Khartoum's statue was in the Square until World War Two, but was removed to another location and never returned to the Square. Changes or indecision regarding monuments demonstrate a similar flexibility of the space to that displayed in the design of the overall Square.

If the statues and monuments within the Square commemorate a national imperial history, the meaning, political power and symbolic sig-nificance of the site cannot be understood without considering its sur-rounding neighbourhood. Geographer Doreen Massey argues that places

---

41   'High Jinks Below Nelson's Column', *The Manchester Guardian* (26 October 1948), 3.
42   Mace, *Trafalgar Square*, 131.
43   Hood, *Trafalgar Square*, 110.

are defined by their links to other places and times, that 'places do not have boundaries in the sense of divisions which frame simple enclosures'.[44] Interpreting Trafalgar Square as porous and connected reveals its links to Victorian Britain and to the Empire it built. Havelock's statue therefore invokes not just mid-nineteenth century London, but also northern India, and the many journeys and relationships between the two since Havelock's death. Just as Driver and Gilbert highlight the imperial landscape that contextualises the Square, Cherry argues that 'Trafalgar Square exemplifies the ways in which a location is defined as much by its surroundings as by its internal reorganizations'.[45] Furthermore, Massey reminds us that even an urban space as apparently well-defined as a square still plays a role within a larger environment; the class and economic aspects of this were implicit in Nash's redesign of the streets leading south to Trafalgar Square in the early nineteenth century.

The political power represented by the buildings and streets within easy reach of the Square is also important, particularly in light of the site's use by political protesters. The Square stands at the head of Whitehall, a street reshaped in the 1860s as 'an architectural manifestation of the institutions of state', that stretches for less than half a kilometre from the Square down to the river Thames and the Houses of Parliament in Westminster.[46] It is also within easy walking distance of the Prime Minister's residence in Downing Street, and to Parliament, both goals for protesters. The proximity of these places has been a concern to authorities seeking to control gatherings in the Square because of fears that protests might spill out. Large gatherings in the Square were also a real concern for the police because of the potential to disrupt traffic in central London and the possibility that protesters might vent their frustrations on the surrounding shops and businesses. This indicates more general anxieties about the London 'crowd' as an entity that needed to be controlled in the interests of public order.

44   Massey, *Places and their Pasts*, 29.
45   Cherry, *Statues*, 671.
46   *Ibid.*, 674.

Early protest

According to London historian Jerry White, the 'excitable' London crowd had a long history of protest, violence and 'disorder for disorder's sake', driven in part by poverty and a lack of other outlets of political expression.[47] In addition, London's vast size meant that enormous numbers of people could be mustered in protest, celebration or to watch a spectacle. For example, in July 1866 during a rally calling for electoral reform, a crowd estimated at 200,000 forced open the gates of Hyde Park, which the police had closed to prevent the gathering. White argues that the crowd was understood as powerful and volatile and that reform leaders feared its unpredictability and potential for violence as much as officials did.[48]

This understanding of the 'crowd', and its association with crime, disorder and the labouring poor, unemployed or working classes, formed part of the social and political context in which the Square was built. The 1830s saw reform of the Parliament and voter enfranchisement that, while expanding the number of men who could vote, still excluded the majority of male voters and all women. According to Clive Bloom, the Reform Bills of 1832 'represented a safety valve against revolution and a widening of participation for those very communities who would have most to lose from a challenge to their economic and *moral* control by the "lower" orders.'[49] In other words, the Reform Bills in 1832 reinforced divisions between the middle and upper classes on the one hand and the poor working classes on the other in the same way that the construction of Regent Street had in the 1820s.

Working class and radical frustration at continued exclusion from formal national participation overflowed in the 1840s in the Chartist Movement. This group staged the first major protest in the new Trafalgar Square on 6 March 1848, when approximately 15,000 Chartists and

---

47   White, *London*, 354.
48   *Ibid.*, 372.
49   Clive Bloom, *Violent London* (Basingstoke: Palgrave Macmillan, 2010), 191.

sympathisers gathered to call for parliamentary and enfranchisement reforms. Five hundred police officers were required to restore order.[50] Some press reports described the tone as mischievous rather than violent, and others characterised the crowd as unemployed or merely curious, 'for the most part the refuse of a crowded city'.[51] The fear of the mob, especially by shopkeepers eager to protect their property, continued for days after this demonstration, and culminated on 13 March 1848 when, in response to a Chartist mass meeting in Kennington Common, police and troops were posted across London in anticipation of revolutionary violence. This affirmed the government's worst fears, and 'Londoners [...] believed themselves on the brink of civil war'.[52] Although the protest slowly dissipated, the potential of the newly completed Trafalgar Square as a central rallying point for protesters had been affirmed, as had the fears of authorities.

One official response to such fears had been the creation of a police force, Home Secretary Robert Peel's Metropolitan Police in 1829. This was strongly resisted by radicals as an imposition on their rights by an unconstitutional armed force. Bloom argues that the creation of a public police force represented a formalisation of a power relationship between the state and the poor that was deeply biased against the people: 'The very concept of *order* embodied in the idea of a *standing* police force was entirely dependent on a new network of power relations and state-led rhetoric opposed to the interests of the ordinary lower orders and the demands of social democrats [...]'.[53] As demonstrated in 1848, Trafalgar Square was one battleground in a larger urban landscape in which this contest was played out. The constant potential for conflict between the state and the people was spatial as much as it was political.

The use of the site by poor or unemployed people to press for government help or recognition continued throughout the late nineteenth century. Several incidents of mob violence in the West End took place in

---

50   Hood, *Trafalgar Square*, 65.
51   Quoted in White, *London in the 19th Century*, 366.
52   White, *London in the 19th Century*, 367.
53   Bloom, *Violent London*, 192.

1886, including a demonstration on 8 February 1886 during which West End shops were looted and their windows broken, and this event, coupled with a dense fog, fed panicked rumours of thousands of marchers descending on London, prompting businesses and banks to close early.[54] By the next summer, socialist protest was occurring regularly in Trafalgar Square. On 8 November 1887, the Commissioner of Police banned meetings in the Square given the growing number of protests and, in response, radical groups demanded 'freedom of speech and the right to protest in London's great open spaces'.[55] Protest organiser William Morris wrote to the *Daily News* arguing against the ban on meeting in the Square on the basis that gatherings there did not obstruct the surrounding thoroughfares and could be easily controlled by police. He accused the Government of 'sit[ting] upon the safety valve' of public discontent by banning meetings in 'the most convenient place in all London for a large open air meeting', and urged readers to attend the 13 November meeting because 'if an impressive protest is not made [...] the liberty of free speech in London is gone, and will have to be slowly and laboriously won back at the cost of great suffering [...] and of abundant inconvenience to the public at large'.[56]

On 13 November, on what came to be called 'Bloody Sunday', a large meeting was called by the Metropolitan Radical Federation, and marchers converged on the Square from working class areas around London. As the demonstrators moved towards the centre of London they clashed with police and broke up into smaller groups. In the Square itself, large numbers of police and troops were ready to prevent marchers from challenging the ban on meetings. According to observer George Smalley: 'Mounted police were in strong numbers at every angle of the square; on the south side a line of policemen "four deep, elbows touching", on the other sides just two deep'.[57] When groups of marchers reached the Square, they were

54   White, *London in the 19th Century*, 376 and Mace, *Trafalgar Square*, 166.
55   White, *London in the 19th Century*, 377.
56   RCH Briggs, 'Morris and Trafalgar Square', *Journal of William Morris Studies* (Winter 1961), 28–31. <http://www.morrissociety.org/publications/JWMS/W61.RCHB. pdf>, accessed 25 March 2013.
57   In White, *London in the 19th Century*, 378.

confronted by police and troops, including some with fixed bayonets, who tried to disperse the crowd. According to Porter, the fracas 'was marked by unusual brutality', and a West End business owner complained to the House of Commons that 'for at least an hour, the most frequented streets in the West End of London [were] entirely at the mercy of the mob'.[58] By the end of the day, around two hundred protesters had to be treated in hospital, and two police officers had reportedly been stabbed. At least two people died in the following weeks as a result of injuries.

The Times was unequivocally supportive of the police response to the protesters, framing the demonstrations as tantamount to a threat to both London and the nation: '[Police Commissioner] Sir Charles Warren receives this morning the congratulations and thanks of the whole law-abiding population of this country', crediting his 'masterful arrangements' for defeating 'the determined attempt made yesterday to place the metropolis at the mercy of a ruffianly mob [...]'.[59] In Parliament, Under Secretary for Home Affairs, Stuart Wortley, pointed out that 'Trafalgar Square had attracted many persons besides politicians, and incitements had been uttered and disturbances occasioned without elucidating any political question, but only causing widespread alarm and extra work to the already hardworked police'.[60] This implied that, despite the violence, at least some understood the incident as an exception to the usual rule of the Square's role as a legitimate location for political protest. 'Bloody Sunday' was also a demonstration of the animosity between the working and propertied classes in late Victorian London, and the use of Trafalgar Square had symbolic value for both sides. The authorities, including the editors of *The Times*, in response to middle-class fears of poverty and the crime and degeneracy associated with it, saw the open space of the Square as a threat, a place where large numbers of people could gather, become agitated by firebrand speeches, and rampage through the surrounding streets, as they had done

---

58   Porter, *London*, 253.
59   'Contents of this day's paper', *The Times* (14 November 1887), 9.
60   'Mr Stuart Wortley on the Trafalgar Square meetings', *The Manchester Guardian* (16 November 1887), 8.

in 1886. In reaction to the violence, Charles Warren, the Commissioner
of Police, re-issued a ban on public meetings and speeches in the Square,
this time including processions in nearby streets on 18 November 1887.[61]

The next weekend, police and newly recruited civilian 'Special
Constables' guarded the Square and other central London places to prevent further attempts at protest. Alfred Linnell, an off-duty law-writer,
went to the Square after a meeting at Hyde Park, got caught up in a police
charge and died of blood poisoning after his thigh was crushed by a police
horse.[62] His death embarrassed the police, and the protesters organised a
massive funeral with a hearse bearing the notice 'Killed in Trafalgar Square'.
A pamphlet was printed for his funeral to raise money for his orphaned
children,[63] and at his funeral well-known socialist and textile designer
William Morris spoke against the 'wanton brutality' of the police.[64]

In a debate in the Commons the next year, even the Home Secretary,
who had supported Warren's ban on the use of the Square for meetings,
agreed that 'it is desirable that popular discontent should find a free and
open voice, and should not be driven to express itself in secret conventi-
cles and other ways'.[65] Here, the Square can be understood as a symbolic
national forum that represented the *right* to express anti-establishment
views, an activity seen by protesters as central to British democracy, and
even accepted by authorities as a necessary outlet for legitimate political
differences. On the other hand, the police ban on meetings was a tactic
intended to control the much larger social issues of poverty, crime and
unemployment, expressed as a 'consideration of public order, the protection of persons and property'.[66] For the less powerful, however, this was
tantamount to a restriction of free speech and political participation. The
contest over the Square in this example was a contest over two different
visions of the nation associated with the extension of the franchise and the

61   See House of Commons Debate, 1 March 1888 vol 322 cc1879–954.
62   White, *London*, 378–9 and Hood, *Trafalgar Square*, 76.
63   Mace, *Trafalgar Square*, 192–193 and Hood, *Trafalgar Square*, 76.
64   'A Trafalgar Square Martyr', *The Manchester Guardian*, 19 December 1887, 6.
65   House of Commons Debate, 1 March 1888 vol 322 cc1879–954.
66   *Ibid*.

question of free speech and its relationship to public order. Protesters used the Square in 1887 to press for a reimagined place in the nation, demanding *political* visibility by means of *spatial* visibility. That the Square was at the heart of a larger official political landscape and was symbolic of imperial power made it the perfect site to demand this recognition, as did its location in a larger economic and social urban environment that felt itself vulnerable to the actions of the mob.

The outright ban against meetings in the Square was to stay in place until October 1892 when new Home Secretary Asquith relaxed it with a set of regulations that limited the times of day in which demonstrations could take place and stipulated that protesters inform the Commissioner of Police four days before the planned event. The next year, in reaction to an Anarchist demonstration in the Square (at which he was burned in effigy), Asquith summed up his approach to allowing regulated demonstrations:

> [...] provided persons assemble together peaceably with an object which is not in itself criminal or in violation of the law – I am not going to interfere with them any more in Trafalgar Square than I should if they held their meeting in some hole [...] to use a vulgar expression, they 'let off the steam' and act as a kind of safety-valve to feelings and opinions which are only dangerous so long as they are held in suppression and are not properly looked after [...][67]

Here, for both the Government and protest organisers, the Square was a public space that should be available for public use, and a valuable platform for venting public opinion. John Parkinson identifies the importance of public space on making public claims, arguing that representative decision-makers pay attention to smaller groups' claims when it appears as if the wider public takes such claims seriously, 'which means that claims need to be made in publicly visible and accessible places' such as Trafalgar Square.[68] However, by controlling space, particularly one with a high profile and a history of public claim-making, the Police Commissioner in 1887 sought to close off this avenue of public expression; in William Morris' words,

---

67 House of Commons Debate, 14 November 1893 vol 18 cc889–890.
68 John R. Parkinson, *Democracy and Public Space: The Physical Sites of Democratic Performance* (Oxford: OUP, 2012), 42.

'sit[ting] upon the safety valve', and potentially forcing protesters towards potentially more violent tactics. By 1892, a new Government had recognised the Square's role as a safety valve for public opinion, although the impact of such opinion on policy-making was less clear.

Control of and access to the Square has been an ongoing point of conflict between government authorities and protesters throughout its history. However, official power had another manifestation that many people were very eager to see: the display of royal pageantry. Trafalgar Square has regularly been an important point along processional routes through central London for other national and imperial events, in part because people could gather there to both watch and participate in national (and imperial) spectacles. In 1852, for example, the funeral procession of the Duke of Wellington, victorious commander at the battle of Waterloo and former Prime Minister, passed Trafalgar Square where crowds had gathered to catch a glimpse of the spectacular funeral car with its 'magnificent dolphins'.[69] Churchill's funeral procession in January 1965 attracted thousands of people, and the route, printed in the *Radio Times* so that the whole nation could follow it in their imaginations, passed Trafalgar Square as it went up Whitehall and turned into the Strand. The Royal Coronations of George V in 1911, George VI in 1937, and Elizabeth II in 1953 also all passed the Square. During the 1937 coronation, spectators tried to catch a glimpse of the coronation procession, but the photographs of the event by Henri Cartier-Bresson also show people sleeping, resting, climbing the lions (and each other), and staring off in different directions.[70] Maps and descriptions of the routes of coronation processions featured in souvenir publications that were designed to help people around the Empire participate vicariously in the events by following the procession through the streets of central London. The 'official souvenir programme' of the coronation of George VI and Queen Elizabeth made this explicit: 'this Programme shall reach British subjects wherever they may be, in city, waste or wilderness. Its pages will enable one and all to participate more easily, in spirit if not

69   Hood, *Trafalgar Square*, 67–68.
70   See Hargreaves, *Trafalgar Square*, 48–49.

in person, at the solemn ceremony which, on May 12th is being enacted in the capital of the empire'.[71] Such publications linked a much larger audience to royal processions than the one actually watching and participating, and embedded London's landmarks, including Trafalgar Square, in official displays of royalty and national identity.

Visitors to the Square have also been drawn there for tourism or leisure. The 1883 Baedeker Guide to 'London and its Environs' encouraged visitors by describing Trafalgar Square as 'one of the finest open places in London and a great centre of attraction'.[72] However, by 1914, Ward and Lock's *Guide to London* was more ambivalent, calling it 'a large open space described by Sir Robert Peel as "the finest site in Europe", though it can hardly be said that the best use has been made of it. One critic has indeed gone so far as to call it "a dreary waste of asphalt with two squirts"'.[73] The Australian High Commission produced several guidebooks for Australians visiting London, and the 1924 and 1930 guides treated Trafalgar Square as 'a conveniently central point' and started several of its suggested day-long itineraries there, although it did not treat it as a destination in itself.[74] The Square also featured in British Railways' domestic advertisement to 'See England by Rail' in the 1950s, as well as in guide books for foreign and Commonwealth servicemen and women during the two World Wars.[75] In the 1970s, Trafalgar Square was the most popular tourist site in London, with twenty-five per cent of visitors originating elsewhere in Britain – the importance of the Square clearly appealed as much to domestic visitors as to foreign ones.[76]

71   *The Coronation of their majesties King George VI and Queen Elizabeth: Official Souvenir Programme*, King George's Jubilee Trust, 1937.

72   Karl Baedeker, *London and its environs, including excursions to Brighton, the Isle of Wight, etc* (London: Dulau and Co, 1883), 137.

73   *Guide to London* (London: Ward, Lock and Co Ltd, 1914), 72.

74   *The Australians' Guide Book to London*, High Commissioner, Australia House (1924) and (1930).

75   See David Gilbert, 'London in all its glory', 279–297.

76   J. Hillman, 'Tourist view of London' *The Guardian* (4 March 1974), 7.

While bye-laws have always prohibited certain business activities, tourism is an approved activity in the Square, and the relationship between tourists and authorities is relatively untroubled. Even climbing atop the lions, which is officially discouraged, is not prohibited in practice. There has been a continuity of this type of use over time, as evidenced by many images of families in the Square covered by pigeons, or waving from the back of one of the lions.[77] A 2005 exhibition in the adjacent National Portrait Gallery linked the photographic history of the Square to visitors' records: 'photographs that record defining moments of history taken in Trafalgar Square have a resonance with British audiences since so many of us have at one time or another been photographed in the Square'.[78]

As well as families taking snapshots, professional photographers have also long been associated with Trafalgar Square. In 1952, for example, the post-war annual gift of a Christmas tree from Norway provided one enter-prising photographer with an irresistible seasonal opportunity for souvenir photographs. 'If we had one child ask us where Santa was we had a dozen', he said, continuing to explain how his wife had made a suit out of inexpen-sive fabric, and with his father-in-law playing the part of Santa, they took to the Square and its tourist customers. Santa's visit, however, was brief, because at the time, setting up a business in the Square was prohibited: 'he was detained by two stern-faced officers who asked him for his name and address – a shattering inquiry considering the tiny auditors in the vicinity'. This enforcement of the letter of the law was not appreciated by bystanders:

> 'What are they doing to Santa?' wailed a shattered and bewildered tot as the little drama was played out to its tragic end among the laconic pigeons. The boy burst into tears and there were not a few among the silent adult watchers who lived again through the agonies of childhood disillusionment.[79]

For some people, however, Trafalgar Square has been a site not just for visiting or running a business, but for living in. In the late nineteenth cen-tury on a nightly basis, for example, the Square hosted a demonstration

---

77 See Hargreaves, *Trafalgar Square*, 68–69.
78 Hargreaves, *Trafalgar Square*, 13.
79 'Short life for Santa Claus in Trafalgar Square: police demand his name and address', *The Manchester Guardian* (24 December 1952), 10.

of the disparate but parallel experiences of life in central London. In the 1880s whole families slept rough there, huddled under newspapers. An image from *The Illustrated London News* on 29 October 1887 shows food being distributed to 'campers' in the Square in the small hours of the night by charity workers, as well-dressed men and women watch from the railings above, their hats and fine clothes a contrast to the rags worn by the people below. The Prime Minister, Lord Salisbury, mentioned this use of the Square in an October 1887 letter to Queen Victoria, when up to four hundred homeless people were recorded in the Square:

> I have just walked through Trafalgar Square. There was no sign of disorder; only about 300 dirty people clustering around the column. The streets were in no way obstructed or disturbed, and everything was going on as usual.[80]

For the sleepers in the Square, the fountains were more than ornamental, and a letter written to the Office of Works in 1887 objected to people sleeping in the Square and then 'performing their ablutions in the morning in the basins of the fountains'.[81] At other times, however, this was a result of celebration rather than poverty. A photograph from 1945, for example, shows a sailor, having slept out in the Square with his two female companions, washing his face in the fountain on Victory in Europe Day.[82] Perhaps the sailor would have thought twice if he had known what else the fountains had been used for over the years. When they were unveiled on 2 May 1844, they were quickly used as public toilets and rubbish bins, and in 1854, cleaners removed the decomposing remains of cats and dogs from the sludge at the bottom of the fountains.[83] Whether as bath, toilet, rubbish bin or crowd control, the fountains symbolise some of the uses that authorities wanted to prevent or control.

Cherry, in examining the archive of material available on the early years of Trafalgar Square, claims that the site has been subject to physical as well as symbolic flexibility and transformation over time:

80 In Mace, *Trafalgar Square*, 176.
81 *Ibid.*, 171.
82 See Hargreaves, *Trafalgar Square*, 56.
83 Hood, *Trafalgar Square*, 57–59.

the intractability of materials, the conflicting accounts, contradictory dates, the inde-
terminancy of the site, the fluidity of its borders and edges [...] an unresolved and
unsettled space, the square has been the subject of reorganisation and re-imagination,
demolition and rebuilding, its space has been drawn and re-drawn in relation to its
surrounding area.[84]

The year 2003 saw a major transformation of the site the northern terrace
was pedestrianised and, although this was broadly welcomed, other changes,
such as the ban on feeding pigeons, were more controversial.[85] Having already
refused to renew the license of the last seed seller, in 2003 the Greater London
Authority (GLA) enacted a Trafalgar Square bye-law to prohibit feeding of
pigeons in the Square, to help make the space cleaner and more amenable
to visitors.[86] Until this time, large flocks of the birds had gathered there
daily to feed, and featured in countless visitors' photographs. According to
Mayor Ken Livingstone, who led the efforts to rid the Square of pigeons, the
regulations would mean 'a cleaner, healthier environment on [sic] Trafalgar
Square. Cleaning costs and treatments have been reduced as guano levels are
now fairly low'.[87] For Livingstone and the GLA, the basis of the anti-pigeon
measures was hygiene and amenity, and this justified the feeding ban, as well
as subsequent bird-of-prey patrols intended to discourage the pigeons from
returning. Livingstone's approach also rested on what, in his view, the Square
symbolised: 'what Trafalgar Square looks like is representative of the city
as a whole. It is a public space for the enjoyment of Londoners and visitors
alike'.[88] However, as with many aspects of the Square, this did not go uncon-

---

84   Cherry, *Statues*, 666.
85   Greater London Authority, 'Trafalgar Square re-opens'. Press release, 2 July 2003.
      <http://www.london.gov.uk/view_press_release. jsp?releaseid=1828 > accessed 5
      December 2007.
86   Greater London Authority, 'New byelaw prohibits pigeon feeding in Trafalgar Square'.
      Press release, 24 October 2003. <http://www.london.gov.uk/view_press_release.
      jsp? releaseid=2032 > accessed 7 September 2009.
87   *Ibid.*
88   Ken Livingstone, 'Ken Livingstone: Why we must remove the pigeons from Trafalgar
      Square', *The Independent* (24 January 2001) <http://www.independent.co.uk/opin-
      ion/ commentators/ken-livingstone-why-we-must-remove-the-pigeons-from-trafal-
      gar-square-703824.html> accessed 7 September 2009.

tested. MP Tony Banks defended the presence of the pigeons, arguing that 'as the pigeons were identified so closely with the square, and thus contributed to the image of one of London's most popular tourist attractions, they should be cherished'.[89] The pigeons (and their absence) show how changes to the Square can be contentious. Nicholas Penny, Director of the National Gallery in 2009, provided another example, sniffily commenting that the Square has become 'uncivilised' since its pedestrianisation in 2003: 'Levels of civil behaviour are incredibly low. As I speak, people are riding the lions and climbing up as far as they can on the reliefs of Nelson's Column'.[90] For artist Mark Wallinger, whose sculpture appeared on the Fourth Plinth in 1999, the question of competing views on the use of the Square rests on one question: 'you have to decide who is the Square for'.[91]

This question has been partially answered by the presence of 'Heritage Wardens' who have the task of assisting visitors, and making sure the Square's bye-laws are obeyed. These uniformed wardens remind people in the Square that regulations about its use must be respected. In official information, they are emphasised as a safety feature: 'They wear distinctive uniforms and are on Trafalgar Square 24 hours a day, 7 days a week. Since they were introduced in 2000, the Heritage Wardens have helped reduce graffiti and have made people feel safer on the square'.[92] The dual role of the Heritage Wardens in providing assistance and surveillance demonstrates the complex relationship between users and the authorities. Sometimes these interactions have been benign or friendly, especially during officially organised events or national celebrations. During the 1937 coronation, for example, police turned an indulgent eye to the celebrations in the Square, 'laugh-

89   Alex Stewart, 'Banks stands up for "gentle London pigeon"' <http://www.independent.co.uk/news/uk/home-news/banks-stands-up-for-gentle-london-pigeon-634544.html> accessed 28 March 2013.

90   Quoted in '"Bloody awful" Trafalgar Square shatters calm of the National Gallery', *The Times* (10 July 2009) <http://entertainment.timesonline.co.uk/tol/arts_and_entertainment/visual_arts/article6676759.ece >accessed 27 July 2009.

91   *Ibid.*

92   Greater London Authority, 'Heritage Wardens', <http://www.london.gov.uk/ trafalgarsquare/visit/wardens.jsp> accessed 8 Sep 2009.

ingly preventing a very good-tempered crowd from climbing up [Nelson's Column]'.[93] At other times, such as 'Bloody Sunday' in 1887, confrontation was violent and destructive.

These relationships hint at how official power informs access to the Square by setting boundaries on the way individuals can use it and create their own narratives within it, a theme explored in more detail through this book. The Heritage Wardens are the most recent manifestation of official control over the space that has included police, armed soldiers, and crowd wardens for specific events since the 1840s. In contemporary daily use of the Square, according to surveys by the GLA in 2009, this control is mostly welcome.[94] As it delineates what is allowed and what is prohibited, however, official control of the Square subtly describes the parameters for contests over national inclusion and belonging. This is because by setting boundaries for the use of the space, the authorities help to determine the way that the nation can be imagined within it. However, these attempts at control have not gone unchallenged, as the following chapters will show.

93   Humphrey Jennings and Charles Madge, eds, *Mass-Observation Day Survey: May 12 1937* (London: Faber and Faber, 1937), 153.
94   Greater London Authority, 'Heritage Wardens'.

Empire, Suffrage and the Great War (1900–1918)

If Trafalgar Square's role in the twentieth century can be characterised as a relationship between the official, imperial and monolithic on the one hand and the subversive, counter-hegemonic and multiple on the other, then its meanings have been constituted by many groups staking their own claim in the space. This chapter will explore how this process unfolded in the years before and during the Great War, a period of less than twenty years. Its main focus is the use of the Square by campaigners for female suffrage and how they activated the history and symbolism of the site to make their claims for expanded participation in national life. These activities serve as a good example of the process of interaction between the 'top-down' and the 'from-below' that has made the Square a unique debating chamber on the nature of British national identity. The chapter also discusses how the Square was used during the Great War, almost entirely for pro-war purposes such as recruiting and fund-raising, and how it also formed part of the backdrop to the Armistice celebrations in November 1918.

This chapter will begin to show how the Square helped to construct national identity in three main ways. The first is through ongoing reference to the imperial history represented by the Square's design and monuments. This was certainly not limited to Trafalgar Square itself, but was an important aspect of the surrounding landscape of central London in the pre-war period, a time when Empire was still central to British society, politics, economics and consumption. Debates about the impact of imperialism on everyday life notwithstanding, there is no question that Britain's economy was an imperial one, and that London was at the centre of this system in many ways. The second is the prominence of Trafalgar Square. Although it is obviously a public space, the Square's location and national visibility helped to raise publicity and bolster the legitimacy of activities that took place there. The Women's Social and Political Union (WSPU) showed the value of the Square in helping an organisation's activities gain national publicity, an achievement they saw as important. This was enhanced by the fact that police approval was required for legal use of the Square, so the possession of such permission can be understood as conferring legitimacy.

The final theme of this chapter is the way that the previous history of the Square's use for protest or resistance helped to support the claims of Edwardian female suffrage protesters. Through a detailed discussion of the WSPU's use of the space, this chapter will show how this process unfolded and how the geographic concentration of meaning, or what others have called 'accreted symbolism',[1] helped to shape their political claims and the spatial strategies they used to achieve them. The history of the Square, both symbolic and experiential, was activated through use, inscribing the Square itself in the narrative of political transformation used by the WSPU. During and after the war, the Square was used very differently, but was still critical to the national narrative being constructed there.

Edwardian London: public life in an imperial landscape

The pre-war period in Britain has been characterised as a relaxed and self-confident 'long, leisurely summer' defined by the consumption of luxuries and free time enjoyed by Britain's wealthier classes. However, that the period also saw rapid change and deep uncertainty in domestic society and politics at a time 'when Victorian liberal values collided with an ambiguous Edwardian modernity expressing itself most dramatically through social reform, constitutional change, industrial militancy and belligerent feminism'.[2] Eric Hobsbawm characterised the forty years before the Great War as a period that saw 'the society and world of bourgeois liberalism

---

1   See Dwyer, *Symbolic accretion*.
2   Alan Simmonds, *Britain and World War One* (London: Routledge, 2012), 5.

advancing towards what has been called its "strange death" as it reaches its apogee, victim of the very contradictions inherent in its advance'.[3] Such changes to the industrial environment, economy and demographics of Britain in general, and London in particular, were accompanied by changes to social structures, with social divisions of class and gender subject to violent contestation.[4] By the beginning of the Great War, power was highly concentrated in the hands of the very few, with ten per cent of the population controlling ninety-two per cent of Britain's wealth. This highlighted one of the central social concerns of progressives at the time: the political rights of the poor and unemployed.[5] Political representation was an important part of radical vision of a 'just society' and the extension of the national franchise in 1832, 1867 and 1884 had enabled most men to vote in parliamentary elections. Local government had also become more representative.[6]

As the biggest city in the world, greater London's approximately six and-a-half million inhabitants were experiencing a profound change in economic structure from industrial manufacturing to finance, as well as an artistic renaissance, becoming more fashionable and creative.[7] Compared to other big cities, such as New York, Paris or Berlin, London was bigger and wealthier. In 1900, London was the biggest city in Europe, one of only five on the continent with more than one million residents, a financial and administrative centre with global reach, and a magnet for migrants from all over Britain and the Empire. London's West End, home to shops, clubs and upper-middle class residences, was an important part of this set of networks, a place for the middle classes to socialise, consume luxury goods and set the political agenda for the nation and the Empire.[8] The central

---

3   Eric Hobsbawm, *The Age of Empire: 1875–1914* (London: Abacus, 1987), 10.
4   David Powell, *The Edwardian Crisis: Britain 1901–1914* (London: Macmillan Press Ltd, 1996).
5   Schama, *A History of Britain*, 316.
6   Powell, *The Edwardian Crisis*, 5.
7   Jose Harris, *Private Lives, Public Spirit: Britain 1870–1914* (London: Penguin Books, 1994), 19–21.
8   Erika Rappaport, *Shopping for Pleasure: Women in the Making of London's West End* (Princeton, New Jersey: Princeton University Press, 2000).

districts of London were undergoing a concentrated building boom, with new government buildings, department stores, hotels and residences mushrooming throughout Whitehall and the West End.[9] All these improvements meant that this area, including the landmarks of Piccadilly Circus, Oxford Street, Leicester Square and Trafalgar Square, enjoyed special significance as national sites of power, consumption and display. What is more, this landscape was peppered with overt reminders of the British Empire, with the symbolism of imperial heroism and victory repeatedly flagged in the statues, monuments and building facades of Edwardian London.[10] At the heart of this symbolic landscape was Trafalgar Square, a site containing symbols and iconography that linked it to British history and global Empire in emphatic terms. Describing London between the wars, W.S. Percy reminds us that 'Trafalgar Square was truly the centre of the Empire. One has only to see the floral offerings from every colony displayed at the foot of Nelson's Column each year on Trafalgar Day to realise that Canada and South Africa have wisely chosen to place their London homes here'.[11]

The urban landscape that provided a setting for the lives of Londoners in 1900 was subject to much lower levels of official control than at any other time later in the century. While the Metropolitan Police was already almost seventy years old, the spectre of 'the crowd' that had menaced respectable citizens in the 1880s, for example, still haunted official concerns over public celebration of the relief of Mafeking and even the Great War Armistice in 1918. This 'outcast London' was still regarded as 'coarse, brutish, drunken and immoral [...] an ominous threat to civilisation', that could erupt unpredictably onto the streets.[12] However, despite large-scale changes to streets and neighbourhoods in the nineteenth century, designed in some cases to restrict access to wealthier areas, metropolitan spaces were enlivened by a diverse mix of people. From the banal activities of work, shopping and transport, to the intermittent spectacle of official events, London's public character before the Great War was varied and engaging. Much of this

9   White, *London in the 20th Century*, 10–12.
10   Schneer, *London 1900*, 19.
11   W.S. Percy, *The Empire Comes Home* (London: Collins, 1937), 24.
12   Gareth Stedman Jones, *Outcast London: A Study in the Relationship Between Classes in Victorian Society* (Oxford: Clarendon Press, 1971), 285.

quotidian mix occurred in public, in the city streets. According to Jerry White, 'Edwardian London was still very much an outdoor city, with as much of the eighteenth century about it as the twentieth [...]. It was an unrepeatable mix of spontaneity and big set-piece events that distinguished London street life in the era of imperial splendor before 1914'.[13] This was to change throughout the twentieth century, as public life went inside and underground, especially during two periods of wartime restrictions of street lighting, alcohol strength and public house opening hours. However, London street life in the Edwardian era was more lively than at any other time in the century to follow.

These two aspects of London in 1900 – the imperial built environment animated by the presence of a rich public life – are perhaps best illustrated by the public celebrations of the relief of Mafeking on 18 and 19 May, 1900, an event that 'seemed to bring onto London's streets every one of its 6.5 million citizens who could crawl from their cots or hobble from their beds'.[14] By the end of the First World War, public display of patriotic fervour could still be taken up enthusiastically by Londoners who flocked to traditional areas of congregation and celebration. Contemporaneous reporting describes the widespread jubilant mood, as in this *Manchester Guardian* report:

> Not for a moment was the purpose of the day left in doubt. People set about the business of festivity with a determination which appealed to one, amid much that was incongruous, as a typically English characteristic [...]. Men, women and children stepped briskly out of their morning trains carrying the Union Jacks in their hands instead of the usual umbrellas, and bent their steps hurriedly towards the Mansion House, Trafalgar Square or the Horse Guards.[15]

The draw towards the streets and towards celebrating in public and with other people, was an important part of how national and imperial identity was expressed, and Trafalgar Square was a central part of the landscape in which this occurred.

13　White, *London in the 20th Century*, 308–309.
14　*Ibid.*
15　'London Revels: Last night's fireworks in Hyde Park', *The Manchester Guardian* (17 November 1918), 6.

Edwardian London: women, place and politics

Changes to London's – and Britain's – economy, transport, built environment and politics during the Edwardian period had wide-ranging and diverse impacts. Class and race remained important social distinctions, and gender roles, although transformed throughout the period, remained restrictive. Women and girls of all classes had more access to education, and middle-class women sought work or professional training in larger numbers, but for the majority of women the range of social roles that they could respectably occupy was still quite narrow.[16] At the turn of the century, modesty and propriety were highly valued aspects of femininity, and sexual identities were largely confined to marriage and reproduction, particularly for the middle classes. Both urbanisation and female political representation, however, challenged the scope of women's roles in Edwardian society. As women, particularly the middle classes, sought to participate more in the life of the city, by going out to shop, work, or play, they challenged existing notions of a 'public woman', a figure who had previously been sexualised and associated with danger or transgression in the Victorian imagination.[17] Judith Walkowitz has shown how women out in public had to cope with 'street impertinences' from 'male pests' who tried to draw women into sexually weighted conversation, particularly in the West End of London, an area bordered by Trafalgar Square. However, Walkowitz also exposes the complexities of these interactions, with her examination of female correspondents to newspapers as maintaining a 'division between "decent women" and "disorderly women of the streets"', that revealed the 'late Victorian distinctions of class and gender to be slippery and emotionally charged'.[18] She points out that street banter with men was not always received as threatening by all women, and that being

---

16   See Hobsbawm, *Age of Empire* and Harris, *Private Lives*, 24.
17   See Rappaport, *Shopping for Pleasure*, and Judith Walkowitz, 'Going Public: Shopping, Street Harassment, and Streetwalking in Late Victorian London', *Representations* 62 (1998), 1–30.
18   Walkowitz, 'Going Public', 12 and 16.

'spoken to' was neither uncommon nor unwelcome for many women out in public. In terms of political participation, however, the role of women of any class was very limited. Although women could be active in government at a local or municipal level through participation in local school boards or some councils, they were still excluded from the national vote.

The campaign for female suffrage, in demanding that women be able to vote on par with men, used tactics that resonated with increased female visibility in public spaces, particularly by middle-class or 'respectable' women. For opponents of female enfranchisement, members of suffrage organisations were unnatural or unfeminine, and this was in part linked to their focus on activities outside the home. Themes of deviant gender roles appeared in anti-suffrage representations of campaigners, portraying them as immature and girlish, remiss in their maternal duties or 'unsexed', hysterical spinsters.[19] Some commentators felt that female education and intellectual life were deleterious to health and reproduction.[20] For those opposed to female suffrage, traditional definitions of femininity and respectability, and suffrage supporters' failure to comply with them, were important rhetorical tools used to discredit suffrage campaigners and their political goals. Furthermore, according to their detractors, female suffrage campaigners were also neglecting their duty to the nation, posing a threat to the nation and the British 'race'. Social Darwinists emphasised women's vital role as mothers, or 'progenitors of the race' in the context of falling fertility rates across the economically advanced countries of Europe.[21] In the UK, official concern was so high that the 1911 census was used for the first time to survey the fertility patterns of the entire country.[22] In the global political environment of the early twentieth century, racial superiority was seen as vital in the competition amongst nations, and enfranchised women appeared to pose a threat to British power because decreased fertility rates were seen by their detractors as linked to women's activities outside the domestic sphere.

19   See Lisa Tickner, *The Spectacle of Women: Imagery of the Suffrage Campaign 1907–1914* (London: Chatto and Windus, 1987).
20   Harris, *Private Lives*, 27.
21   Tickner, *The Spectacle of Women*, 186.
22   Simon Szreter, *Fertility, Class and Gender in Britain, 1860–1940* (Cambridge: Cambridge University Press, 1996), 2.

The relationship between traditional Edwardian femininity and national identity was a visible part of the campaign for female suffrage. Spectacular events such as public pageants, designed to raise publicity and widen support for female suffrage, depicted women as visible and loyal members of the British nation and Empire. However, for middle-class Edwardian women, appearing in public for the purposes of protest or demonstration was both unusual and personally difficult because of the social expectation that women's place was primarily in the private, domestic sphere. As a result, large numbers of women protesting in public had a strong impact as both a tactic and a spectacle.[23] As the campaign for female suffrage developed, processions and mass meetings were used by members of several groups to raise public awareness of the issue and thereby pressure the government into formally considering female enfranchisement in Parliament.

The Women's Coronation Procession of 1911 provides one example of national and imperial identity in pre-war London in which the suffrage movement participated enthusiastically. In 1911, Britain celebrated the coronation of a new monarch, George V, and on 17 June, as part of the celebrations, female suffrage groups organised a procession with participants from Britain and overseas. The national, imperial and international display wound through the streets of central London and included marchers dressed as powerful women from the past to emphasise women's contribution to the nation. The event was well-attended and good weather added to the spectacle. As with similar national events, Trafalgar Square was full of people watching the procession, as it was 'a favourite vantage point [... with] people on all the hundreds of seats erected for the coronation.'[24] The leaders of the WSPU, co-organisers of the procession, had discussed the possibility of militant protest action during the Coronation, a time when 'the Empire was to be in London' and 'the Suffragettes held the

23   Tickner, *The Spectacle of Women*.
24   *Ibid.*, 125.

centre of the world stage'.[25] They decided instead to participate peacefully in the procession and to express their loyalty to the nation in this message to the newly-crowned monarch:

> Our [WSPU] tender to their Royal and Imperial majesties, the King and Queen, their loyal and devoted service [...]. May the Empire prosper under their guidance and advance in strength, in honour and in righteousness, and may men and women, rendered equal before the law, secure, by their united endeavour, a future greater than heretofore.[26]

This message demonstrates how the WSPU constructed its political aims in terms of national and imperial identity by presenting its members as loyal to the monarch, nation and Empire while also seeking to redefine the polity to include women as voting participants.[27] The coronation procession also showed that the WSPU was aware of the publicity that such spectacle could bring, and used it as a means to pursue political goals. This had precedent in an Edwardian penchant for public display and pageantry in general, and the use of such tactics by the wider female suffrage movement in particular. Between 1907 and 1911, for example, four large suffrage processions took place, including the Women's Coronation Procession, organised by the National Union of Women's Suffrage Societies (NUWSS), WSPU, and others. These events made use of urban place and rhetorical and visual propaganda, and relied upon the media and local crowds for publicity. According to Jorgensen-Earp: 'Just as Parliament greeted bills concerning women and children with raucous laughter and rude jokes, the press responded to the suffrage movement with what amounted to a press blackout. The need to overcome this blackout formed the first major

---

25 Pankhurst, Christabel, *Unshackled: The Story of How We Won the Vote* (London: Hutchinson, 1959), 175.

26 Pankhurst, *Unshackled*, 184.

27 For more on the role of imperial or transnational identities in the suffrage struggle, especially as they overlap with national ones, see Ian Christopher Fletcher, Laura E. Nym Mayhall and Philippa Levine, eds, *Women's Suffrage in the British Empire* (London and New York: Routledge, 2000).

rhetorical exigency leading to an increase in WSPU militant tactics'.[28] As a political strategy, public spectacle helped the suffrage movement gain coverage in mainstream newspapers which had been largely uninterested in the campaign, and through the newspapers, the public.[29] Central London provided landmarks familiar to the British public that formed the perfect stage for their campaign of visibility.[30]

In 1908, two processions took place that combined the visual effect of a uniform colour scheme with the spectacle of large numbers of women in public. The use of colours, such as the WSPU's purple, white and green, by women when campaigning, in processions, or simply to show support for female suffrage, had a strong visual effect that resonated with conventional notions of femininity. Recognising the business potential, shops kept stocks of clothes and accessories in the WSPU's three colours, 'all with the purchasing power of middle-class Suffragettes in mind'.[31] This included items such as 'walking skirts', and accessories made purposefully for processions, with the intent that 'this [political] movement becomes identified in the mind of the onlooker with colour, gay sounds, movement, beauty'.[32] One of the first processions to use these visual methods to spectacular effect was on Saturday, 13 June 1908, when the NUWSS organised a large procession through the streets of central London to a rally at Albert Hall that passed Trafalgar Square. Newspaper reports of the events highlighted the impressive visual aspects of the procession, drawing out the dignity of the participants:

28   Cheryl Jorgensen-Earp, ed., *Speeches and Trials of the Militant Suffragettes: The Women's Social and Political Union, 1903–1918* (London: Associated University Presses, 1999), 20.
29   Tickner, *The Spectacle of Women*, 58.
30   Katherine Kelly, 'Seeing Through Spectacles: The Woman Suffrage Movement and London Newspapers, 1906–1913', *European Journal of Women's Studies* 11/3 (2004), 331.
31   Diane Atkinson, *Suffragettes in Purple, White and Green: London 1906–1914* (London: Museum of London, 1992), 19.
32   *The Suffragette*, quoted in Atkinson, *Suffragettes*, 15.

> Marshalled in a mighty phalanx over ten thousand strong and nearly two miles in length, the women suffragists marched through the streets of London on Saturday [13 June 1908] to the Albert Hall – and the conquest of the vote. The event was in every way a triumph. It was impressive and picturesque.[33]

Newspaper coverage of the 13 June procession hints at the event's popular reception, as well as attitudes to the question of female suffrage. It also points out the ongoing challenges that many women must have faced in participating:

> many of the sightseers that had come to scoff remain to cheer [...]. Here and there, of course, a ribald jest or a vulgar music-hall refrain was heard; but they were few and far between and altogether missed fire in the presence of the dignified attitude of the women.[34]

This report suggests some of the difficulties that many women must have felt in challenging social and political norms through their participation in such female suffrage marches. It also indicates the novelty of female processions and their value as spectacle, as well as how participants were using urban space in new, untried ways, at least for middle-class women.

The organisation of the 1908 march, as well as its size and visual impact, was a focus of comment in both public and private documents. NUWSS organiser Phillipa Strachey received messages from participants congratulating her on the march, in which the importance of positive publicity is evident. Edith Palliser, for example, wrote: 'I feel sure the reports in the newspapers will amply reward you, for the procession accomplished what we have long desired [...] to show that an earnest desire does exist among the "quiet" women.'[35] The next weekend, on 21 June, the WSPU organised a 'Women's Sunday' along similar lines, which was reported in the sympathetic *Daily News* in visual and spatial terms that were specifically feminised:

33 'Women's Day', *The Daily News* (15 June 1908), 7.
34 *Ibid.*
35 Palliser, Edith 'Card dated 19th June 1908 to Miss P. Strachey from Edith Palliser, from Amsterdam congratulating her on the success of the Procession of June 13, 1908'. Microfiche Box 1, Vol 1 (A–J), June 1908, 9/01/0407 (ALC/407). The Women's Library, London Metropolitan University.

It was a white demonstration, touched with the green and purple that have become the emblem of the Women's Social and Political Union, which organised it. More than two-thirds of the women taking part wore white dresses, giving to the streets of London through which they marched under their silken banners a richness and refinement of colour such as the grandest of military pageants has never supplied.[36]

The decision to mount such processions in central London helped to raise publicity and gain media coverage, positioning the campaigners firmly in the public sphere. Furthermore, this material shows how female participation subtly began to redefine public femininity in new ways, and also points to the use of London itself to demand changes to national policies. These demonstrations helped to create a narrative of a visible, politicised femininity that challenged mainstream Edwardian expectations of women as restricted to the private domestic realm. However, these processions were highly controlled and organised, and their use of the streets had been approved by the authorities, as demonstrated by the police that accompanied the processions. As one participant reported, 'Escorted by the mounted police we trudged along. People cheered as we passed along and handkerchiefs waved.'[37] While participants in this and other such events may have expressed discomfort at being publicly 'on display', these marchers largely conformed to gender expectations of respectability.

Reactions to women in public, however, were not always as positive as this. In a violent conflict on what came to be known as 'Black Friday' on 18 November 1910, police confronted suffrage protesters as they approached the Houses of Parliament with the intention of presenting a petition to Prime Minister Asquith. Some of the violence was sexual, with reports of women being 'seized by the breasts' or having their undergarments torn off.[38] One of the petitioners described her experience:

---

36   'Beauty of the banners', *The Daily News* (22 June 1908), 7.
37   E.G. Murray, *Personal diary, 1895–1918* (Murray, Utah: Family Heritage Publishers, 2007), 114.
38   June Purvis and Sandra Stanley Holton, *Votes For Women* (New York, Routledge, 2000), 139.

[...] we had to run the gauntlet of organised gangs of policemen in plain clothes, dressed like roughs, who nearly squeezed the breath out of our bodies [...] women were thrown from policemen in uniform to policemen in plain clothes, literally till they fainted. A lady told me a policeman had told her he would kick her when he got her down – and he did.[39]

The sexual nature of the police violence was a harsh reminder of the gendered nature of urban public space in this period, with women still subject to officially sanctioned brutality at the hands of police. Over one-hundred women were arrested for attempting to access the Houses of Parliament on Black Friday, and the incident is an extreme example of protesters' real vulnerability that found a more mild expression in discomfort at being 'in full view' when marching in public.

The novelty of the spectacle of women marching in public gradually wore off, and the 1911 coronation was the last large procession as a part of the campaign for female suffrage. By this time, the WSPU had begun using more violent tactics, in part as a result of frustration with government inaction and in an attempt to maintain publicity. Over the next three years, these would grow to include window-breaking, arson and property damage.[40] While the WSPU continued to attract large numbers of supporters in public displays, its 'militant' tactics were a violent and disorderly contrast to the more traditional image of attractive and orderly femininity depicted in the processions discussed above.

The WSPU in Trafalgar Square

The themes touched on so far – political representation, gender roles, national identity, imperial landscapes and the use of public place – all converged in the WSPU's use of Trafalgar Square to make its political

---

39 Hertha Ayrton quoted in Purvis and Stanley Holton, *Votes For Women*, 139.
40 See Melanie Phillips, *The Ascent of Woman: A History of the Suffragette Movement and the Ideas Behind It* (London: Abacus, 2003).

demands. Although it was neither the only group that campaigned for female suffrage, nor the only militant group, this chapter's focus on the WSPU is due to the well-known spatial aspects of its political campaign, and in particular its use of Trafalgar Square. Known as 'Suffragettes', the WSPU's militancy was at odds with the strategies of other groups, which favoured less violent or publicly visible 'constitutionalist' tactics such as lobbying key politicians or peaceful processions. The WSPU used Trafalgar Square and other parts of London to increase the physical visibility of women and, through the publicity that this tactic raised, the political prominence of their cause. Furthermore, as this section will show, these activities were contextualised with a narrative of liberty and rights drawn in part from the history of Trafalgar Square itself.

From 1906, the WSPU began using militant tactics as a means to pressure government to follow through on a series of parliamentary measures to advance female suffrage. An important strategy in the period from 1906 to 1914 was civil disobedience, summed up by the slogan 'Deeds not Words'. The WSPU used language and techniques that were new to the suffrage campaign in order to make its demands, and 'the union's shocking tactics, courage, fundraising and organisational capabilities set it apart from its predecessors and contemporaries'.[41] These included crowded and sometimes violent rallies in Trafalgar Square that led to marches on the Prime Minister's residence in Downing Street, arrests, and sensational news coverage.

As discussed, much of Trafalgar Square's value for protesters lay in its public visibility, and according to Kelly, for the Suffragettes, 'visibility [was] the cornerstone of their representational strategy'.[42] The central WSPU office was near many newspapers' offices in Fleet Street, and following an earlier shift from Manchester to London, WSPU leader Christabel Pankhurst described the organisation's new choice of location for their headquarters in terms of its value for raising publicity: the location 'was highly convenient for the newspapers who were ever interested in the

41 Atkinson, *Suffragettes*, 51.
42 Kelly, *Seeing through Spectacles*, 329.

militant movement'.[43] Further publicity for their cause was created by the WSPU's methods of causing disruption in public places, the purpose of which was to challenge 'public apathy and press indifference'.[44] Inasmuch as its tactics of property damage and the arrest of its leaders kept the organisation in the newspapers, it was successful in creating publicity for the movement. According to Kelly, this use of spectacle was a central tactic for the WSPU, as it 'laid the groundwork for [political change] by integrating women into the physical landscape and media representations of the city'.[45] Emmeline Pankhurst summed up the importance of publicity in a 1913 speech she gave in Hartford, Connecticut. Comparing the WSPU to a fitful, crying baby, she asserted:

> You have to make more noise than anybody else, you have to make yourself more obtrusive than anybody else, you have to fill all the papers more than anybody else, in fact you have to be there all the time and see that they do not snow you under.[46]

For the WSPU, urban public place was an important stage for such activities, and there were several examples of such uses of the Square between 1906 and 1913. On 19 May 1906, 'the first large open-air women's suffrage gathering ever to take place in London' was held in Trafalgar Square, although it was mostly attended by men.[47] The meeting was intended to raise publicity to pressure the Government and at the meeting Sylvia Pankhurst, daughter of WSPU leader Emmeline Pankhurst, appeared to understand her campaign for the rights of women as part of a lineage of previous counter-hegemonic uses of the space, including previous contests over the right to vote. According to Kelly, in 1906 the WSPU understood the Square as a site anchored in a narrative of political rights and used the space to legitimise its campaign by linking itself with political movements of the past. In other words, it mobilised the Square's history as a site of counter-hegemonic protest to frame its own radical demands:

43   Pankhurst, *Unshackled*, 70.
44   Tickner, *The Spectacle of Women*, 9.
45   Kelly, *Seeing through Spectacles*, 330.
46   Emmeline Pankhurst, 'Freedom or death', 13 November 1913, <http://www.guardian.co.uk/theguardian/2007/apr/27/greatspeeches/print> accessed 4 April 2013.
47   June Purvis, *Emmeline Pankhurst: A Biography* (London: Routledge, 2002), 83.

> [WSPU leaders] selected Trafalgar Square as a meeting place for their first march [in
> 1906] to reinforce the WSPU's symbolic alliance with radical political disturbances
> of earlier decades [...]. As an icon of open-air radical agitation, free speech and urban
> confrontation, Trafalgar Square provided an ideal platform for the WSPU slogan,
> 'Deeds not Words'.[48]

If the Square's historical resonance was important, it was also significant as a highly visible site. In 1906, publicity was a new weapon of the WSPU, and while its daily business was being conducted in semi-public meetings or in its offices, Trafalgar Square was a venue to make arguments and rally support as publicly as possible.[49] Devices such as banners and bands were used, and prominent speakers addressed the crowd. At the May 1906 meeting, for example, the radical newspaper *Labour Leader* reported that Emmeline Pankhurst and Labour MP Keir Hardie spoke at 'an imposing demonstration, led by three bands, with hundreds of banners flying in the breeze'.[50] Dwelling on the visual and aural spectacle that a large number of people in the Square presented, the report emphasised the physical occupation of the space as a significant aspect of the demonstration: 'a vast crowd thronged Trafalgar Square, surrounding Nelson's Column [...] the stronger the argument the louder the vast assembly cheered'.[51] The call to action was repeated by Keir Hardie, who urged 'all women to carry on a persistent agitation until the degrading political outlawry from which they suffer is ended and they have attained their rights as citizens'.[52] For these supporters, direct action was the best way to pressure the government to adopt the policy of female suffrage, and the Square was important because it provided a historically significant platform from which to broadcast the message of the 'natural rights' of citizens as widely as possible.

Two years later, Parliamentary consideration of a female suffrage bill was refused, this time by new Liberal Prime Minister Asquith. In response, the WSPU organised a 'mass meeting' in Trafalgar Square on Sunday, 11 October 1908, where leaders Emmeline and Christabel Pankhurst and Flora

48   Kelly, *Seeing through Spectacles*, 351.
49   Jorgensen-Earp, *Speeches*, 19.
50   *Labour Leader* (25 May 1906), 10.
51   *Ibid.*
52   *Labour Leader* (25 May 1906), 11.

Drummond addressed the crowd from the plinth of Nelson's Column.[53] Thousands of handbills were distributed urging supporters to 'Help the Suffragettes to Rush the House of Commons on Tuesday Evening, 13th October, 1908 at 7.30'.[54] Emmeline Pankhurst's exhortation was reported the next day:

> [...] I believe that the men and the women who go to see [the protesters] demand political justice will help them to reach the people's House of Commons – the House which belongs to the women as well as the men of this country.[55]

At this rally, the WSPU's leaders described the space of the House of Commons as both material and metaphorical, a building only a few hundred metres away that was the political possession of all members of the nation. Pankhurst connected the physical occupation of Parliament to the right for women to be recognised within the nation; physical visibility in public places such the House of Commons symbolised national visibility. However, they were accused of inciting the crowd to illegal activity, and Emmeline, Christabel and other WSPU members were arrested after the Trafalgar Square meeting as a result of this metaphorical claim to national space. Responding to the Pankhursts' exhortations, Eunice Murray, a visitor from Dunbartonshire, Scotland, participated in the 13 October 'rush' on the House of Commons. She had travelled to London to assist in the suffrage campaign and had attended processions in London the previous June, as well as the October protest:

> The W.S.P.U. have had a scene in Parliament Square, such a performance on the part of the Government truly laughable. All London ablaze with excitement because a few women were going to the House, shouting crowds blocking the way, mounted police patrolling the streets and charging the crowds. Rows and rows of police to protect the legislators, a double row on each side of Parliament Street down towards Charing Cross – 8,000 police out, not to protect these gallant men from armed mobs, but to stop a few women coming to ask for their rights.[56]

---

53   Jorgensen-Earp, *Speeches*, 25.
54   Purvis, *Emmeline*, 113.
55   'Yesterday's Mass Meeting', *The Daily News* (12 Oct 1908), 9.
56   Murray, *Personal Diary*, 119.

In this account, Murray highlights a discrepancy between the female protesters and the numbers of police sent to control them. The control of and
right to use public space was a perennial theme for suffrage campaigners,
pointing to the importance of the contestation of space in national political struggles. For example, according to Emmeline Pankhurst, at the 11
October rally 'police were there [...] taking ample notes of our speeches.
We had not failed to notice that they were watching us daily, dogging our
footsteps, and showing in numerous ways that they were under orders to
keep track of all our movements'.[57] In addition to surveillance, the police
repeatedly responded to WSPU rallies or movement through the city by
erecting barriers, often with their own bodies. Control over the use of public
places demonstrated official attempts to suppress a national narrative that
included enfranchised women.

Five years later, in 1913, in reaction to the campaign of militant Suffragettes and concerned about the possibility of civil disobedience as a result
of very large gatherings, the Government removed the right to protest in
Hyde Park. This was an attempt to control the size of gatherings and was
aimed at discouraging the 'monster meetings' in the Park that had been
one aspect of the female suffrage campaign. However, Trafalgar Square still
allowed a smaller number of protesters to make a noticeable demonstration, and on 4 May campaigners responded to the Hyde Park ban with a
meeting in the Square 'to defend free speech'.[58] According to sympathetic
press coverage, restrictions to the 'right' to gather in particular public places
were tantamount to an abrogation of public liberty. This was reflected in
one of the resolutions put to the crowd: 'That this meeting declares the
unalterable right of free meeting and free speech, regarding them as the
only safeguards of public liberty and of the claims of minorities'.[59] Material
from other speeches provides additional evidence that suffrage campaigners understood 'liberty' in this way. According to police records of the 4
May 1913 meeting, speakers began by linking British national identity with

57   Purvis, *Emmeline*, 113.
58   *The Suffragette* (9 May 1913), 498.
59   *Ibid.*

a historical notion of liberty that included the right to gather in public places. One early speaker used history to reinforce the point, mentioning an 1887 Trafalgar Square rally:

> I do not believe any liberty loving British Citizen wishes to leave it to the Police in this Country to decide whether it is right or safe that people should be allowed to hold a meeting in Hyde Park. We won the right of meetings in Hyde Park in 1867 just as we won them here in 1887 [...].[60]

As discussed in Chapter 1, in 1887 meetings were banned in the Square given the growing number of protests, and when radical groups challenged this, there was a violent confrontation with police and troops that resulted in injuries and death. In 1913, these events were well known to speakers at the rally, and the police transcriptions record multiple uses of this place-history to contextualise the protesters' demands. As one unidentified speaker said, 'They [the Government] are going to have the lesson of 1887 again, if they are not careful' and 'What was necessary in 1887 is necessary now'.[61] Keir Hardie, Labour leader and a longstanding supporter of female suffrage, was even more explicit in his rhetorical use of Trafalgar Square to link the 1913 event to past struggles for popular rights: 'We have had the finest demonstration seen in Trafalgar Square since 1894. On that occasion the demonstration compelled the Liberal Government to give the eight hour day'.[62] Police transcripts also show that Hardie linked the Square with labour solidarity and political recognition, and alluded to past struggles with authorities that had occurred there:

> Give nobody any excuse for using violence [...]. Remember 1887. Do not forget friends that our strength lies in our solidarity. Solidarity in Trafalgar Square [...] so as to bring about the day when our own class shall have become the ruling class, and thereby brought assurance of liberty, equality, and fraternity.[63]

60 MEPO 2/1556, Trafalgar Square Meetings: 'Free Speech Defence Committee' (1913), 2. Available in the National Archives.
61 *Ibid.*
62 MEPO 2/1556, 13.
63 *Ibid.*

Finally, female suffrage campaigner Charlotte Despard, leader of the
Women's Freedom League, made a strong claim for women to be heard
within the public sphere: 'They are trying to stop women from speaking,
but the woman [sic] are going to speak and act [...]. The capitalistic papers
will say that there was [sic] a few women in Trafalgar Square today. The great
heart of the men of country [sic] – and the women – is beating in unison.'[64]
In her speech, Mrs Despard connects female visibility – and audibility – with
national unity in support of free speech rights. Referring to the authorities'
opposition to both their cause and their use of public space to protest, she
describes the gathering as both powerful and counter-hegemonic. Here,
the contest over the right to gather and be visible in Trafalgar Square symbolised the right to national visibility in the form of enfranchisement.
The Square also acted as a highly visible stage for a spectacular gathering
designed to publicise the cause. *The Suffragette* coverage of the 4 May 1913
rally, for example, emphasised the visibility of the event by describing the
large number of attendees, painting a picture of a massive public gathering:

> For two hours before the meeting began an enormous crowd thronged the square.
> Processions of men and women poured down from all quarters of London, until
> a mass of 30,000 people, all eager to declare their dissent from the suppression of
> Suffrage meetings in Hyde Park, were gathered round the Nelson Column.[65]

The nature of the spectacle, however, went beyond the large numbers of
participants. While *The Times* emphasised the size of the crowd, which it
estimated at 20,000 to 30,000, it was the 'riotous' behaviour of the crowd,
rather than its size, that provided the main spectacle:

> During a 'free speech' demonstration in Trafalgar Square yesterday, intended as a
> protest against the prohibition of suffragist meetings, there were scenes of great dis-
> order, and the police were compelled to take vigorous measures to disperse the crowd
> [...] free fights ensued between the police and the crowd [...] several [police] helmets
> rolled off on the melee and scores of people were knocked down and trampled upon.[66]

64   *Ibid.*
65   *The Suffragette* (9 May 1913), 498.
66   'Scenes of Disorder in Trafalgar Square', *The Times* (5 May 1913), 8.

This line of reporting continued across a range of editorial positions. The *Daily Citizen* reported that 'There was literally a fight for free speech in Trafalgar Square yesterday afternoon.'[67] The pro-suffrage *Daily News* described the 'exciting scenes in Trafalgar-square', blaming the trouble on an 'unauthorised' speaker who, in attempting to speak from the southern side of the plinth of Nelson's Column, had caused a conflict with police and 'brought a disorderly finish to an otherwise orderly and successful meeting'.[68]

Later in the year, on Sunday, 10 August 1913, another rally took place in Trafalgar Square, attended by WSPU's Sylvia Pankhurst, who had been temporarily released from prison as a result of her hunger strike. A large crowd gathered in the Square to protest against the imprisonment of George Lansbury, a prominent supporter of female suffrage, and the 'monster demonstration' was described in *The Suffragette* newspaper as 'one of the largest held in the Square in recent years, some estimates giving the number of the audience as 20,000'.[69] At the rally, a fight with police broke out, and Pankhurst was one of several people arrested. In the meetings on 4 May and 10 August 1913, the themes of liberty and rights were closely linked to the spatial occupation of Trafalgar Square as a tactic by which to demand political inclusion. The banner and speeches urging participants to 'Remember 1887' and the description of the Square as a battlefield for free speech reference notions of citizens' liberty and the right to national visibility that were central to the WSPU's political claims.

The WSPU used Trafalgar Square to bring its national narrative to public attention through two main devices: public visibility and the operationalisation of radical history. First, the WSPU's strategy of feminine visibility in public manipulated the generally masculinised character of the 'street' to press for national legal recognition alongside physical recognition. As they campaigned for female suffrage, they negotiated the tension between raising publicity for their cause and maintaining the respectability

---

67 Quoted in *The Suffragette* (9 May 1913), 498.
68 'Free Speech Demand', *The Daily News* (4 May 1913), 1.
69 *The Suffragette* (15 August 1913), 768.

that defined middle-class Edwardian femininity. This public visibility, however, was central to the suffrage movement's claim to national participation. According to John and Eustance, 'the campaign for female suffrage helped to change the very idea of what constitutes the "political nation", but being part of a wider women's movement they [were] also central to the disintegration and reformation of ideas around the "public and private".'[70] For the Suffragettes, the struggle over space, and the right to be present in it, represented a struggle over the right to be literally counted as part of the national body politic. As they worked for formal political inclusion in the nation, they made themselves more visible members of it through their use of central London, and their goal of national *political* visibility was pursued through a strategy of *spatial* visibility in urban places, notably Trafalgar Square.

In doing so, the WSPU drew on the site's meaning in two specific ways. First, Trafalgar Square created access to a place-based historical narrative of citizens' rights, such as those of free speech and assembly, which the WSPU used to frame and legitimise their political demands. Trafalgar Square's place-history, such as the 1887 protests, reinforced the WSPUs political claims by linking contemporary struggles to the past. Second, the evidence of contests over the use of the Square, including how they were understood by the WSPU leadership, the police and other national authorities, suggests that conflict over Trafalgar Square represented a struggle for national recognition, and that the occupation of the Square, for example through a rally or demonstration, functioned as a bid for national belonging. These aspects of the Square's use – the activation of place-history in aid of contemporary political goals and the occupation of metaphorical national space – are two of the contributions that Trafalgar Square makes to how Britishness was and is constructed.

70   Angela John and Claire Eustance, eds, *The Men's Share? Masculinities, Male Support and Women's Suffrage in Britain, 1890–1920* (London and New York: Routledge, 1997), xv.

The Great War

By the outbreak of the Great War, despite the opposition to the British government that had animated the WSPU throughout the early twentieth century, its leaders Emmeline and Christabel Pankhurst suspended their militant activities and turned their efforts to supporting the war. This included a 'patriotic demonstration of the WSPU' in Trafalgar Square at which Emmeline expressed confidence in Prime Minister Lloyd George and pride in the amount of money raised in public subscriptions to the War Loan.[71] The Pankhursts also changed the name of their newspaper to *Britannia*, a periodical that supported war against Germany accompanied by military conscription and the internment of alien nationals in Britain.[72] Even though this appeared to be a change from their position on the legitimacy of the British state before the war, it was broadly in line with the public mood, particularly in 1914 at the beginning of the conflict, and was underpinned by the view that the British system, despite its faults, was still superior to Germany's.

The Pankhursts' change of direction resonated with the views of many Londoners, who gathered in crowds in Trafalgar Square, Whitehall and outside Buckingham Palace chanting 'we want war' on 4 August 1914.[73] Lurid rumours about German barbarity against 'gallant little Belgium' fed this popular mood, even though the complex system of alliances which brought Britain into the war was complicated and difficult to understand. *The Times* gave its version of public reaction:

> Towards midnight Trafalgar-square, Whitehall, Parliament-street, and Parliament-square were filled with a solid mass of people [...]. A profound silence fell upon the crowd just before midnight. Then as the first stroke rang out from the Clock Tower, a

---

71   'Faith in the Premier, Mrs Pankhurst's praise of Lloyd George', *The Observer* (18 February 1917), 9.

72   Phillips, *The Ascent of Woman*, 294; and Robb, *British Culture*, 37.

73   Simmonds, *Britain*, 35.

vast cheer burst out and echoed and re-echoed for nearly 20 minutes [...] by nightfall
[the next day] Whitehall and Parliament-street, Trafalgar-square, and the Mall were
packed with cheering masses.[74]

However, just as it saw an overflowing of public enthusiasm, Trafalgar
Square had also seen an anti-war protest in the days immediately preced-
ing Britain's declaration of war on Germany. Reported in *The Manchester
Guardian* as 'the largest Trafalgar Square demonstration for many years', on
2 August 1914, orators including Labour MP Keir Hardie and Charlotte
Despard denounced the war on the basis of British support for Russia.[75]
According to the report, despite the presence of 'a few youths waving Union
Jacks from 'buses', the meeting enjoyed 'many thousands of supporters', who
'overflowed into Whitehall and some distance up the Strand'. *The Times*
reported the rally somewhat differently, suggesting that the crowd in the
Square was 'such as one generally finds at Sunday afternoon meetings at
this spot', and was made up of 'the poorer section of the population' and
'many foreigners [...] the great majority [of whom] were Germans'.[76] It also
reported that the 'Socialist speakers' struggled to make themselves heard,
their voices drowned out by hecklers in the crowd, and when a red flag
was raised, 'The rejoinder came at once. A Union Jack was waved and the
cheering and singing of patriotic songs grew louder than ever'. As had happened many times before, control or occupation of the Square represented
a contest over the national narrative.

Support for the war, however, was not necessarily as uniformly enthusi-
astic or eager as *The Times* would have its readers believe. Gregory, for example, suggests that the crowds in London were made up of more 'interested
spectators' than actual supporters of war;[77] as with many other instances
of crowd activity in London, including the celebration of VE Day in 1945
discussed in Chapter 4, the spectacle was in part the throng itself, with
people joining the crowd in part for the excitement and diversion of it.

74 'London and the Coming of War', *The Times* (5 August 1914), 9.
75 'Anti-war demonstration in London', *The Manchester Guardian* (3 August 1914), 10.
76 'War Protest Meeting', *The Times* (3 August 1914).
77 Adrian Gregory, *The Last Great War: British Society and the First World War*
    (Cambridge: Cambridge University Press, 2008).

Gregory's analysis of press coverage of the event also reveals a class element, suggesting that 'middle-class youths' were responsible for heckling the speakers and disrupting the otherwise unanimously resolved meeting. He concludes that 'pro-interventionist sentiment was almost certainly a minority opinion as late as 2 August, and possibly until the actual declaration of war on 4 August'.[78] In terms of Trafalgar Square, this event and the response to it underscores the tension between official propriety and radical protest that has often characterised the use of the site.

Despite the likely complexity of the public mood, the use of the Square for anti-war protests from 1914 to 1918 was limited. In part this was a result of the Defence of the Realm Act (DORA), which imposed a number of wartime regulations including restrictions on the sale of alcohol, night-time blackout and attempts to control public 'disturbances' caused by anti-war publications or meetings. In line with DORA, in October 1914, the First Commissioner of Works changed the existing regulations to make it more difficult for groups to use the Square for repeated meetings.[79] An example was a 1916 Easter Sunday anti-war demonstration in the Square planned by a group of Labour and anti-conscription organisations, which was prevented by the police on the basis that such a rally could cause a major disturbance. In an extended discussion in the House of Commons, Home Secretary Herbert Samuel defended the government's denial of the use of the Square for anti-war protesters. He argued that permission had been denied because participants intended to speak out against conscription, and that this could result in illegality. The meeting was ostensibly denied approval because of potential law-breaking, not because the Government disagreed with the thrust of the meeting itself. Samuel emphasised the role of the Square as a site in which free speech, a precious British right, could be exercised, expressing the view that 'it would not be proper for the Executive, even when engaged in a vital struggle such as this, to prevent the expression of opinions on the part of members of the public that were hostile to the policy of the Government in power'.[80]

78   Gregory, *The Last Great War*, 15–16.
79   Mace, *Trafalgar Square*, 208.
80   House of Commons Debate, 1 June 1916 vol 82 cc 2953–3050.

Free speech notwithstanding, Samuel was also on record as describing such anti-conscription views as 'highly mischievous and most unpatriotic', a danger to the national interest, and potentially discouraging to British troops and their allies.[81] His argument that the reason to shut down the meeting was because it might encourage illegality (rather than because the Government disagreed with the message of the protesters) might be disingenuous. It is certainly reminiscent of the official reaction to the Suffragettes encouraging supporters to 'Rush the House' in 1908, which was to lead to the prosecution of Emmeline Pankhurst and others for 'incitement'. In terms of the use of the Square, even though the Government appeared to value it as a site of 'free speech', it still managed to control what happened there when it strongly objected to the content of that free speech. Shutting down meetings was not the only way to control activities in the Square. Mace describes detailed planning in 1916 by the Office of Works to construct a building on the site which would occupy almost the entire Square, making it impossible to hold public meetings, with the assumption that any demonstrations would be radical or anti-government in nature. Despite support from Cabinet and the view of the Commissioner of the Metropolitan Police as 'in the public interest', this scheme was not pursued.[82] This episode shows, however, the importance that authorities felt Trafalgar Square had as a site at which to express dissident political views, and the degree to which they equated control of an iconic public space with control of public opinion.

Despite evidence of official disquiet, anti-war demonstrations were not commonplace during the war. In fact, the response to a call for volunteers to join the British Army overwhelmed some recruitment offices, and August and September 1914 saw literally hundreds of thousands enlisting in a massive 'rush to the colours'.[83] Given its central location, Trafalgar Square was used throughout the war for recruiting for a range of purposes:

81   *Ibid.*
82   Mace, *Trafalgar Square*, 210.
83   Simmonds, *Britain*, 41.

Trafalgar Day [22 October 1915] was mainly a recruiting rally [...]. There were recruiters on every plinth shouting a summons to duty. The wreaths stacked against the column were half-hidden by rows of soldiers, some of them wounded, all ready to make their appeal. There were also, as the afternoon went on, a group of fresh recruits there, attracted out of the crowd by the straight business talk of the recruiters.[84]

The activities in the Square were not limited to recruiting for troops. A 1917 'Women's Army' recruitment drive aimed to recruit 10,000 women a month for the Women's Army Auxiliary Corps, and the WAAC had a special hut in Trafalgar Square 'where all information will be given'.[85] Other women's organisations also used the Square. In March 1918, the Women's Land Army used the prominence of the site to call for 12,000 new recruits, with the Ministry of Food mobilising the combined spectacle of 'banners indicating the different branches of farm work carried out by women', a hay wagon, and '100 girls assembled in the square to inform recruits about the open-air life which these girls find so attractive'.[86] This activity continued throughout the last months of the war, and even as late as July 1918, the Women's Branch of the Food Production Department was able to report to War Cabinet that, despite 'the wet weather prevailing in London' at the time, it had recruited thirty-four women on a Saturday in Trafalgar Square.[87] Such activities highlighted the role of women in the war effort in the heart of the capital.

The Square was also important as a site of public display and fundraising during the war. A program of entertainments in 1917 showed footage of soldiers and food production, interspersed with official speeches and statements, repeated throughout the day.[88] Also in 1917, a Trafalgar Square 'tank bank' was set up to raise funds by selling War Bonds and War Savings Certificates. Opened by the Mayor of Westminster in November 1917, it was

84   'Crowds flock to Nelson's Column to commemorate the Battle of Trafalgar', *The Manchester Guardian* (22 October 1915) <http://www.guardian.co.uk/theguardian/2012/oct/22/trafalgar-day-first-world-war-archive-1915> accessed 25 October 2012.

85   'War Bonds From a Tank', *The Times* (26 November 1917).

86   'Women's Land Army Rally', *The Times* (19 March 1918).

87   War Cabinet minutes (17 July 1918), National Archives CAB/24/58, 5.

88   Hood, *Trafalgar Square*, 83.

intended 'to lead a vigorous offensive against the enemy', and newspaper reports list the names of several eminent people who were among the first to invest.[89] They also vividly describe the transformation of the Square, hinting at visitors' experiences during this period:

> The whole of Trafalgar-square has been turned into an advertising centre for the War Bond Campaign. Great posters around the Nelson column appeal to people to buy a bond at the Tank, and on hoardings which temporarily hide the fountains there is a display of pictures [...][90]

Captured German artillery pieces and other material, including a field kitchen, were also displayed in the Square to raise public morale, and in the first week, the Trafalgar Square tank bank had raised just over £319,000.[91] Its popularity steadily increased and by the end of the fortnight's campaign in December 1917, it had raised more than £3,400,000, in part aided by a carnivalesque atmosphere in the Square created by singing performers and large crowds.[92] The Square was transformed even more dramatically as part of a subsequent campaign to sell War Bonds during a 'feed the guns' week in October 1918, when a mock-up of a French village was installed, complete with ruined windmill, destroyed buildings and muddy fields. The report in *The Times* begins by assuring the reader of authenticity of the 'battle scene':

> It depicts a ruined farmhouse, riddled with shot and shell holes, and the garden torn up and irretrievably damaged. The building stands where the ornamental fountain used to play, and is surrounded by sandbagged trenches. The Gordon statue is entirely hidden in a ruined church tower [...] and standing on the basin of the west fountain is a windmill, without sails. About 20,000 sandbags have been used to make emplacements and trenches, along which the visitors will walk to feed the guns with Bonds and Certificates.[93]

89 'War Bonds From a Tank', *The Times* (26 November 1917), 5.
90 'Tank Bank Opened', *The Times* (27 November 1917), 5.
91 'Tank Bank's Best Day', *The Times* (1 December 1917), 5.
92 'The Tank Bank's Triumph', *The Times* (10 December 1917).
93 'Feed the Guns', *The Times* (7 October 1918), 5.

Here, fundraising for the war effort was literally constructed in terms of the experience at the front, and transplanted to the heart of London where it acted as a spectacular reminder of the cost of the war. Such radical transformations of the Square for display and fundraising demonstrate both the valuable publicity lent by the location and the way a national narrative was constructed that linked urban space to a foreign theatre of war. The spatial tactic of bringing such a display to Trafalgar Square activated its central public location, and also subtly reinforced the existing martial narrative implicit in its monuments and history. However, it was not radical or counter-hegemonic in the same way that the WSPU's earlier use of the site had been, and was an example of the official and highly controlled use of the Square which has frequently been evident.

When peace was announced on 11 November 1918, the city streets filled with people celebrating. The overall mood was one of relieved jubilation: 'The country seemed mad with relief and joy. Street lamps were lit for the first time since 1915. Dancing became a public craze; couples copulated in doorways'.[94] A *British Pathé* film of the day in Trafalgar Square depicts a packed and heaving space filled with waving and cheering crowds.[95] According to *The Times*:

> Nowhere were they more numerous than before the Royal Exchange and in Trafalgar-square. Round Nelson's Column and the parapets they climbed and held on [...] by their teeth [...]. London with a great moment to celebrate abandoned the hope of suitable words and made festival by ringing of handbells, the hooting of motors, the screaming of whistles, the rattling of tin-trays, and the banging of anything that could be banged.[96]

Paul Ward's analysis of the Armistice celebrations suggests that the reaction was neither uniform nor universal, and begins with Vera Brittain's description of Armistice Day:

94   Simmonds, *Britain*, 283.
95   *British Pathé*, Armistice Day 1918 <(http://www.britishpathe.com/video/armistice-day-1918)> accessed 4 April 2013.
96   'Cheering Crowds', *The Times* (12 November 1918), 10.

'When the sound of victorious guns burst over London at 11am on November 11th, 1918, the men and women who looked incredulously into each other's faces did not cry jubilantly: "We've won the War!" They only said: "The War is over."' She described the dominant emotion as one of relief rather than of celebration of British victory. Yet Brittain is in turn contradicted by C.S. Peel. She described the crowd in Trafalgar Square roaring, 'Have we won the war?' and being answered 'Yes, we've won the war'. A sense of relief was undoubtedly a major factor in the celebrations [...][97]

These accounts both capture a sense of jubilation that can only be expressed by making noise together with many other people, while still expressing the complexity of individual reactions to the end of the War. All of them, however, describe the reaction to the Armistice in terms of how people responded *together*, and this sense of the collective, affectual and even physical is an important part of how the Square worked to construct national identity subtly through participation significant public moments. The colourful and lively display of the Suffragette's banners, the crowds of 'interested spectators' at the outbreak of war and the noisy throng at its close all make real the national imagined community within the Square and the larger imperial environment that surrounds it.

This description above of London's reaction to the announcement of the Armistice is from Monday, 11 November 1918. By the next weekend, the Government response had caught up with the spontaneous celebrations, and 'London, which for the past five evenings had had to content itself with letting off little crackers and coloured lights in Trafalgar-square, decided to go to the party'.[98] The official 'party' in Hyde Park was the culmination of a week in which saw reports of drunkenness and rioting during the celebrations, with 'bonfires in Trafalgar Square [...] German guns destroyed and the plinth of Nelson Column [...] damaged'.[99] This was also reported to the War Cabinet in a meeting on 14 November. In a tone of faint alarm, Chamberlain reported that 'Bonfires had been lit in Trafalgar Square, in

97 Paul Ward, 'Women of Britain Say Go': Women's Patriotism in the First World War, *Twentieth Century British History* 12/1 (2001), 42.

98 'London Revels: Last night's fireworks in Hyde Park', *The Manchester Guardian* (17 November 1918), 3.

99 'Our London Correspondence', *The Manchester Guardian* (19 November 1918), 6.

which German guns had been burned'.[100] In an unwitting reference to its imperial setting, the Prime Minister also noted that Australians had played a 'leading part in increasing the rowdiness of the crowds', and the meeting discussed the importance of getting the colonial soldiers out of London as soon as possible.[101] However, police reports later indicated that celebrations had died down by 18 November, when the crowds in Trafalgar Square were large but orderly, 'with the exception of certain Colonial soldiers'[102] who, according to one newspaper report 'somehow succeed not only getting the diluted liquor of the period, but in getting tipsy on it'.[103] This method of enjoying the Square is probably still going on today.

Overall, Trafalgar Square in the period from 1900 to 1918 was unabashedly imperial and public. Among other things, it was a site from which to watch royal processions and raise money for the war effort in a built environment that celebrated a masculine and martial version of British heroism. It was also one of protest and resistance to official narratives, both in support of female suffrage and against participation in the war. During its demonstrations, the WSPU and its supporters reminded Londoners explicitly that the site had a robust history of use for such protests, and activated the history and memory of these events to buttress its own claims for an expanded franchise for women. In the WSPU's version of Trafalgar Square's history, it had been a space of contest over free speech, liberty and the future of the nation, and previous contests over access to the space and the 'right' to use it to promote popular political causes contextualised their approach to the site. Both before and during the war, Trafalgar Square was also used to protest against participation (including at least one meeting for which official permission was denied) and to raise both morale and funds for the war effort. Here, it was the Square's prominence in the national and metropolitan imagination, and the landscape of central London, that made it a powerful location. This was also the case for the celebrations of

100   War Cabinet minutes (14 November 1918), National Archives CAB/23/8.
101   *Ibid.*
102   War cabinet memorandum from Commissioner of Police Macready (18 November 1918), National Archives CAB/24/70.
103   'Our London Correspondence', *The Manchester Guardian* (19 November 1918), 6.

the end of the war, although the experience of the vital, crowded Square was less ideological than it was physical and affectual.

Overall, the contrasts in this period in the use of the Square begin to reveal how the variety of uses of the Square have made it attractive as a means to frame new versions of national identity. Different groups have been able to draw on aspects of the site – both its history of use and representations of it in its built environment – to help construct different versions of the nation. For example, the WSPU showed how being visible in the Square was part of a political strategy to transform female political enfranchisement, and through this, the nation. Officials, such as the Metropolitan Police and the Commissioner of Works, tried to control these uses by restricting access to the Square itself, emphasising its importance as a symbolic national site while also linking this to the metropolitan impacts on transport or traffic in the heart of London. Finally, concerns about class, the 'uncontrolled' London mob, and potential associated illegality, also animated official responses to the Square's use. The notion of the Square as an unruly and complex debating chamber of the nation, subject to a wide range of uses by many different groups with a national vision, that had characterised the Square throughout its history, would continue between the wars and into the 1940s.

CHAPTER 4

Illuminations (1919–1945)

At nine o'clock on the evening of 8 May 1945, King George VI announced to radio listeners across Britain that the Second World War had ended in Europe. The day had been a public holiday, and like other people across the country, Londoners celebrated in the streets of their city, gravitating towards the centre, seeking out public celebrations. George Broomhead was among them:

> I caught the train to London and made my way to Trafalgar Square and after a while I climbed onto the lion and finished up perched on its head and someone passed me the Union Jack [...]. I was trying to conduct the singing in the crowds at the same time! Those were unforgettable scenes, dancing and singing – it went on all night.[1]

Public celebrations such as these were widely reported in the national and international media, and have become some of the central images of VE Day in Britain. This version of the event is still prominent in the contemporary British imagination, and Trafalgar Square is an important part of this picture. However, the Square in 1945 was the centre of a city that was perhaps less imperial and more focused on domestic matters than before the war, with the bomb damage inescapably evident in London's built environment and in the lives and health of its residents. Whereas Britain at the outbreak of the First World War had been at the height of its imperial reach,[2] by the end of the Second World War only thirty years later, the effects of two wars, and the financial depression and social change they helped drive, saw notions of Britishness unsettled and refashioned.

---

1   Quoted in Royal British Legion, 'Memories of VE Day' <http://www.britishlegion.
     org.uk/ remembrance/ve65/memories-of-ve-day> accessed 27 April 2010.
2   Francis Sheppard, *London: A History* (Oxford: OUP, 1998), 321.

Between 1900 and 1918, the Square can be understood as largely an imperial site with some radical uses by groups calling for new versions of British citizenship, such as the WSPU. By the period between the wars, the site had a firmly established reputation as a place where challenges to the British state, politics and social structures could be made and disputed. As with the London unemployed, the Suffragettes and other radical groups, activities in the Square between the wars often took the form of minority groups using visibility in the space of the Square to make a claim for national visibility, often along class lines. This was interspersed with moments of royal, imperial or national celebration such as the coronation of George VI and the celebration of VE Day in 1945. This juxtaposition of uses – the official and radical, powerful and counter-hegemonic – has often characterised the Square. A 1921 example from the *Illustrated London News* made this point:

> There is a curious inconsistency in the public use of Trafalgar Square, with all its monuments of great deeds in British history. While on Trafalgar Day the Nelson Column is treated as a patriotic shrine, on other occasions the plinth is chosen as a platform for the dissemination of all manner of sedition and disloyalty [during a Sinn Fein demonstration]. Public opinion seems to regard such desecration with apathy.[3]

This chapter focuses on the period from the end of the First World War to the end of the Second. The most detailed analysis concerns the Victory in Europe Day celebrations in 1945, and the role of the Square in helping to frame this event. Sonya Rose argues that the 'ideological work' of nationhood is 'especially trenchant in generating emotional attachment to the nation' in wartime; Trafalgar Square certainly played such a role during these celebrations.[4] However, the period between the wars also saw counter-hegemonic public displays in the Square that were constructed as British, and these formed an important part of the story of the site in the twentieth century.

3   'Contrasts in Trafalgar Square: Desecration and Veneration', *The Illustrated London News* (29 October 1921), 561.
4   Sonya Rose, *Which People's War? National Identity and Citizenship in Britain 1939– 1945* (Oxford: OUP, 2003), 13.

By 1945, the experience of London was framed by a new sense of the fragility of the bomb-wrecked urban landscape, and uncertainty about the extent and cost of the rebuilding effort to come. Trafalgar Square, along with other London landmarks such as St Paul's Cathedral and Buckingham Palace, seemed to embody the combination of resilience, survival and (more equivocally) vulnerability that was evident in first-hand accounts of the illumination of the Square and other sites as part of the VE Day celebrations. Reports of 'crowd-feeling' also point to the emotional and affectual geographies that animated uses of and movements through central London in May 1945. As argued below, events in the Square throughout the period from 1919–1945 bear out the gradual shift in focus of the national imagination from the imperial and international to the domestic and British.[5]

The Interwar Square

There were three types of public event in the Square in the interwar years: major royal events that cleaved to the imperial traditions of display and pageantry and that brought enormous crowds to London's centre; rallies by members of disempowered groups demanding better treatment or conditions, such as the unemployed or the disabled; and meetings held by radical political groups using the Square in part for its history as a site of anti-government protest. For all these groups, the Square provided a history of use or a representative symbolism that both framed their goals and underscored Trafalgar Square's role as a site in which spatial visibility acted as a claim on national visibility.

5   See Ward, *Britishness*.

Royal and imperial events

By the 1930s, with Britain's Empire at its greatest extent, imperial symbolism was particularly evident in central London:

> This triangular area – with Buckingham Palace, Trafalgar Square, and the Houses of
> Parliament at its three corners and the ceremonial routes of the Mall and Whitehall
> along two sides – is probably the most celebrated of all sites of 'imperial' London.[6]

A 1932 poster for the London Underground, illustrated with exciting exotic animals and vernacular imperial architecture, invites the public transport user to 'Visit the Empire by London's Underground', linking different Tube stops to imperial sites such as Australia House (Temple), India House (Aldwych) and the Imperial Institute (South Kensington). Here, the Empire was an important part of how London's geography and built environment were both represented and understood.

Other accounts, however, suggest that the role that Empire played in the interwar and wartime public imagination differed amongst British people. For example, an author for the Left Book Club wrote in 1945 that the Empire had 'little hold on [people's] imaginations',[7] and Colonial Office surveys found evidence of widespread ignorance of the Empire just after the war.[8] Porter argues that the Empire was not very important to people in Britain 'for most of the time that [Britain] was acquiring and ruling the greatest Empire ever', and that its subsequent representations of the late nineteenth and early twentieth centuries have had Empire 'inserted' as culturally significant.[9] Even so, in 1937 W.S. Percy asked 'What makes every Colonial look forward to a trip home?', and placed London, 'The Heart of Empire', at the centre of his answer. Trafalgar Square 'is truly the centre of Empire', but is also important for practical reasons for Colonial visitors:

6   Driver and Gilbert, 'Heart of Empire?', 17.
7   Alexander Campbell, *It's Your Empire* (London: Left Book Club, 1945), 7.
8   Mackenzie, *Imperialism*, 7.
9   Porter, *Absent-minded Imperialists*, 3.

[...] there is one spot in London where every visitor eventually finds himself; that is the short street leading from Trafalgar Square to the Church of St Clement Danes where are the homes of the colonial offices of the different parts of the Empire.[10]

Here, the Square and the streets surrounding it are both the metaphorical centre and a practical one, with the adjoining Strand full of colonial offices. According to Percy, 'you have but to stroll along its pavements to meet men to whom the words Camloops, Calgary, Ronderbosche, Kaitangata, Hokitika, Murrumbidgee and Wagga-Wagga are familiar names, and not simply unintelligible jargon'.[11] The Square also featured in the Empire Marketing Board's 'monster drive' encouraging people to 'buy British', which included 'produce of the Empire overseas'.[12] In what was reported as 'England's Biggest Sign', 'letters twenty feet high blaze the slogan [Buy British]' from the wall of South Africa House, which was under construction at the time.[13]

This environment, however, was enlivened as much by significant and popular national occasions as by advertising or 'the largest letters in England'. Several events in the 1930s showed off this imperial landscape. In 1935, for example, the Empire celebrated twenty-five years of George V's reign with a Silver Jubilee service of thanksgiving at St Paul's. The processional route was crowded from the early morning, with many people having spent the night on the streets. Good weather likely increased the size of the crowd, and Trafalgar Square was full of 'the London behind the front which never sees anything but yet goes to these Imperial occasions primed with hope' that they may catch a glimpse of the spectacle.[14] The King died a few months later in January 1936, and his funeral procession passed Trafalgar Square in its way from King's Cross Station to Westminster Abbey, where his body lay in state. In *The Times* coverage, the weather

---

10   Percy, *The Empire*, 23.

11   *Ibid.*, 25.

12   'Buy British, Great National Effort, Monster Drive', *The Advocate* (Burnie, Tasmania) (22 December 1931), 1.

13   'Our London Correspondence', *The Manchester Guardian* (17 November 1931), 8.

14   'Behind the Front in London's Cavalcade', *The Manchester Guardian* (7 May 1935), 15.

enhanced the solemn urban setting: 'In Trafalgar Square [...] a grey mist was relieved by wintry sunshine only at Admiralty Arch, where it made a lighted background against which fluttered the White Ensign flown at half mast [...] the impressive scene as the procession appeared at Charing Cross was enriched by the last rays of the setting sun'.[15] The elegiac atmosphere was encapsulated by the symbols of Empire – the fluttering ensign, Admiralty Arch and the Square itself – through which the procession moved.

After the death of his father and the subsequent abdication of his brother, Edward VIII, the May 1937 coronation of George VI was a welcome distraction and display of imperial splendour, and dignitaries from across the Empire joined the carriage procession through the streets of London. In a detailed report entitled 'History in Streets and Buildings: the Processional Route', *The Times* claimed that 'it would be hard to find a route of similar length so representative of all the activities of the country and the Empire'.[16] Pointing out that Trafalgar Square was little more than a century old (making it contemporaneous with many other landmarks in central London), the item quoted Sir Robert Peel's view that it was the 'the finest site in Europe'.[17]

The Square itself, on the night of 11 May 1937, had become 'a vast open-air dormitory' with people sleeping out in order to secure a good vantage point for the next day's procession:

> At 5.30 [the night before the procession] about 50 people had climbed the plinth of the Nelson Column, in Trafalgar Square, and were settling themselves for a long wait on the wonderful coign of vantage [... but] the police had their duty to perform, and the people were courteously and sympathetically asked to return to the pavement. Soon the whole length of kerbstone in front of the Column was occupied [...] by 8.30 all the best positions in Trafalgar Square had gone.[18]

15 'A Capital in Mourning', *The Times* (24 January 1936), 12.
16 'History in Streets and Buildings – the Processional Route', *The Times* (11 May 1937), 46.
17 *Ibid.*
18 'The Coronation Vigil – All-night Crowds – Cheerfulness and Good Humour', *The Times* (13 May 1937).

Trafalgar Square was one of the most crowded sites, with people dancing there at 2.30 am the night before the Coronation to music on loudspeakers.[19] By 5.00 am on the day of the event, the trains were packed and more people were heading for Trafalgar Square, doing their best to avoid police barriers along the way, only to find that 'the bottom side of the Square and the Nelson Monument [were] solid with people'.[20] Despite the rain and chill, people continued to celebrate through the night, and in Trafalgar Square, 'two policemen on Nelson's monument were laughingly preventing a very good-natured crowd from climbing up'.[21] According to *The Times*, 'The noise and mirth were at their greatest volume between 1 and 2 o'clock in Piccadilly and at the Circus, Mall and Trafalgar Square'.[22] These accounts hint at the feeling of being in a crowd that seems to be an important part of such events, with the spectacle as much the crowd as the royal event itself. The title of one report on George V's Silver Jubilee summed it up: 'A Crowd "View" from Trafalgar Square – Spectators who saw nothing – But "Would not Have Missed It!"'[23]

David Cannadine points out that such events remained very popular throughout the twentieth century, despite earlier skepticism about the longevity of royal ritual, arguing that by the reign of George VI, 'the monarchy appeared, particularly on grand, ceremonial occasions, as the embodiment of consensus, stability and community'.[24] Like his father, George VI presented himself as a faithful and devoted monarch who refused to leave London during the Second World War, and who was central to national rituals 'in which the royal family, individual families and the national family

19   'In London This Morning', *The Manchester Guardian* (12 May 1937), 7.

20   Jennings and Madge, *Mass Observation*, 104–105.

21   *Ibid.*

22   'Coronation Day', *The Times* (12 May 1937).

23   'Behind the Front', *The Manchester Guardian*, 15.

24   David Cannadine, 'The Context, Performance and Meaning of Ritual: The British Monarchy and the "Invention of Tradition", c. 1820–1977' in Eric Hobsbawm and Terence Ranger, eds, *The Invention of Tradition* (Cambridge: Cambridge University Press, 1983), 140.

were all conflated'.[25] The backdrop to this conflation of the personal and individual with the royal and imperial were the streets of central London, themselves rich in imperial symbolism and representative of the material links to the wealth, products and people across the Empire. This landscape, including Trafalgar Square, acted as a stage for the imperial display of royal events.

The symbolism of these events was unabashedly imperial and they were understood as such by contemporaries, as demonstrated by the media coverage. However, this period also saw a slow transformation from imperial city to a national one as the Empire reached its peak and began slowly to decline, even though the monarchy retained its importance as a symbol of stability and continuity.[26] Alongside the Royal events that passed Trafalgar Square, and Percy's splendid descriptions of the imperial landscape of the streets around it, demands for better government recognition of and support for the disenfranchised, unemployed or disabled were also being made.

Voices of the disempowered

The social changes that occurred as a result of the Great War transformed the role of women, and by 1918, some women had been granted the vote. Although not given the same franchise as men until 1928, many women over thirty were able to participate in politics in new ways after the war. In a newspaper essay published a few days before one of the final rallies in support of an equal franchise for women, Vera Brittain argued that the reason that women still demonstrated in favour of an equal vote was related to being recognised as 'fully human' and as able participants in the public life of the nation: '[The protester's] goal is the work of citizenship which awaits her as soon as she is allowed to play her full part in the

25   *Ibid.*
26   See Cannadine, *The British Monarchy*, 156; and Ward, *Britishness*, 31–36.

making of civilisation'.[27] Forty-two pro-suffrage organisations staged a joint demonstration in Trafalgar Square in mid-July 1927, and *The Manchester Guardian* coverage adopted a tone of fond familiarity, identifying many of the speakers as 'veterans' of the cause. This approach hinted at a movement whose cause was almost won, with none of the emphasis on the spectacle of women in public that had informed coverage of female suffrage protests twenty years earlier. However, the Square remained important to female suffrage campaigners, at least until 1928 when the franchise was expanded.

Much more common in the interwar period was the use of the Square by groups seeking to claim space for themselves within the state and who used the Square to speak directly to government, particularly along lines of social division such as class or employment.[28] The National Union of Police and Prison Officers went on strike in 1918 and 1919, and included rallies at prominent locations around London, including Trafalgar Square.[29] In July 1920, a newspaper report entitled 'labour disputes in London' covered several meetings and protests in London by watermen, railway clerks, gasworkers, textile employees and ex-servicemen, all within a few days of each other.[30] The use of the Square for labour-related protests continued throughout the 1920s: civil servants began a meeting 'punctually at two o'clock' demanding better wages in 1922,[31] and female teachers rallying for equal pay in 1924 were compared to 'old-time suffrage demonstrations' with their 'lettered banners displayed on the plinth', an explicit reference to the Square's symbolic role as a protest site.[32] The groups who used the Square

---

27 Vera Brittain, 'Political Demonstrations: Why Women Still Hold Them', *The Manchester Guardian* (14 July 1927), 8.

28 Trafalgar Square was only one such site it Britain. Glasgow's George Square, for example, also saw protests by unemployed Clydeside workers, notably in 'the Battle of George Square' in January 1919. However, given its history, prominence and location, the Square was the most 'national' of protest sites with events there closer to government decision-makers and national media outlets.

29 Lindsey German and John Rees, *A People's History of London* (London: Verso, 2012), 184.

30 'Labour Disputes in London', *The Manchester Guardian* (5 July 1920), 11.

31 'Civil Servants in Trafalgar Square', *The Manchester Guardian* (3 September 1922), 13.

32 'Equal Pay for Equal Work', *The Manchester Guardian* (12 May 1924), 10.

for protests and rallies throughout the 1920s treated it as a national space, rather than one reserved for Londoners. For example, in 1922, a 'farewell' and fundraising meeting was held in the Square to raise money to support a group of thirty unemployed marchers for their return trip to Birmingham,[33] and in 1927 a group of coal miners concluded a march designed specifically to raise publicity and highlight the 'suffering and tragedy of the workers' homes in South Wales'.[34]

One of the largest rallies of the unemployed in the Square took place on 30 October 1932, when 15,000 mainly peaceful protesters took part in a 'well-arranged, inconspicuously controlled demonstration'.[35] This emphasis on order in the report in a left-of-centre newspaper is likely a response to violence at previous rallies and a concern that the unions not be labelled as irresponsible or violent. By contrast, the discussion in the House of Commons reveals that several arrests, searches for weapons – which uncovered a handful of homemade items, and one person with 'stones in his pockets' – and some scuffles with police took place.[36] Coverage in *The Times* focused on the trials of the speakers arrested at the protest, discussing the charge of an 'attempt to cause disaffection' among the police attending the protest because of a reference to the police union protest of 1919.[37] Just as with the WSPU, the history of the use of the Square contributed to its impact as a protest site.

The Square's use by unions or labour organisations was well-established before this period, with many previous examples. Somewhat less familiar, although not unknown, were marches that culminated in Trafalgar Square by groups such as the unemployed and 'hungry' or the disabled. Often the

33   'Birmingham Unemployed March to London – Farewell Demonstration in Trafalgar Square', *The Manchester Guardian* (28 August 1922), 8.
34   'Marching Miners – Final Demonstration in Trafalgar Square', *The Manchester Guardian* (28 November 1927), 10.
35   'Peaceful Demonstration in Trafalgar Square', *The Manchester Guardian* (31 October 1932), 16.
36   House of Commons Debate, 31 October 1932 Vol 269 Col 1442–1444.
37   'Home News – Marchers' Leader Arrested – Trafalgar Square Speech – Remand at Bow Street', *The Times* (2 November 1932).

end of a longer publicity raising campaign, these events linked the site to a domestic audience beyond London, emphasising the Square's role as a space in which national visibility could be sought. Post-1945 campaigners, notably the Campaign for Nuclear Disarmament's annual marches from Aldermaston to London in the late 1950s and early 1960s, would adopt a similar tactic,[38] but in was in the late 1920s to mid 1930s that such practices repeatedly occurred in the Square. The Hunger March by the unemployed men of Jarrow is perhaps the best known of these events, but these marchers never went to Trafalgar Square. According to Perry, the Jarrow marchers have been confused with a group of Welsh miners who rallied in Trafalgar Square at the same time that the Jarrow men were in London.[39]

Other national hunger marches, however, did end in Trafalgar Square in 1929, 1932 and 1934, and these were covered in press reports that emphasised how far the marchers had come and the hardships they had faced on the route, such as bad winter weather and illness. Here, the Square is the culmination of a journey from all corners of the country; in one 1929 report, marchers came from 'Scotland, Wales, Durham and Tyneside, from Yorkshire and Lancashire, as from far west as Plymouth'.[40] Marches by the blind to Trafalgar Square in 1929 and 1936 also used their presence in the Square to activate claims for national recognition, in this case for better care and improvements in government payments. However, the efficacy of these events in raising publicity is difficult to judge, with one 1932 demonstration dismissed as unenthusiastic: 'Rain did not damp the ardour of the speakers and marchers packed on the plinth itself, but the audience below gradually became smaller as the demonstration dragged on and the organisers

38　The first such March from the CND was on Easter 1958, and went from London to Aldermaston. The CND leadership, however, recognised the value of finishing the march in Trafalgar Square with a big rally, and so changed the route to end in London. This is discussed in detail in Chapter 5.

39　Matt Perry, *The Jarrow Crusade: Protest and Legend* (Sunderland: University of Sunderland Press, 2005).

40　'Hunger Marchers in London', *The Manchester Guardian* (25 February 1929), 14.

had to call repeatedly for cheers to keep up a semblance of enthusiasm.'[41] This coverage in the less sympathetic press is typical of reports that also emphasised the unruliness of the crowd or the moderating actions of police.

One function of 'nationality', according to David Miller, is to highlight the obligations of the role of the state in recognising and caring for the disempowered, that 'the practice of citizenship properly includes redistributive elements of the kind we commonly find in contemporary states'.[42] In this sense, Trafalgar Square was activated as an explicitly *national* site, where groups sought to shape the parameters of their relationship with the state – resolutions calling for better entitlements for the unemployed, for example, show how the Square acted as a platform in which national identity was defined in terms of the right to shared national resources. Most of the unemployed 'hunger' marchers came from outside London, and seemed to regard the Square as a good place to highlight the plight of people living outside the metropole, and therefore further from the gaze of politicians and the mainstream media. That the Square also had a history of being used by other groups for similar purposes only reinforced the protesters' demands, while its prominence helped to ensure they became literally more visible nationally. Overall, such demonstrations were vital for disempowered minority groups, as explained by Storm Jameson, Amabel Williams-Ellis and Vera Brittain in 1932: 'The most important point about the recent demonstrations and hunger marchers is [...] [that] the unemployed who have the most serious complaint are the least articulate. Their way of saying what they want to say is taken from them if it is made impossible for them to demonstrate or to hold meetings or to state their case directly.'[43] Trafalgar Square was one of the places where such statements could be directly made.

41   'The Means Test – Trafalgar Square Demonstration – A Police Charge', *The Times* (31 October 1932), 16.
42   David Miller, *On Nationality* (Oxford: OUP, 1995), 72.
43   Quoted in Juliet Gardiner, *The Thirties: An Intimate History* (London: HarperPress, 2010), 160.

Radical politics

Paul Ward has argued that 'the constraint of ideas and language about what constituted national identity between the wars played a significant part in the restriction of the growth of radical parties'.[44] If such groups did not manage to embed themselves deeply in British political culture, they certainly tried, and Trafalgar Square was one site used by various radical political groups to gain national recognition and support, also bringing them into conflict with each other and the police. Like the tactics used by the Suffragettes before them, the control or temporary 'ownership' of the Square appears to have been a goal in the larger battle of ideas, particularly between Labour or Communist groups on the one hand, and right-wing or fascist organisations on the other. That control of urban space was an important tactic more generally was apparent in the fights that took place over space throughout London during the 1920s and 1930s.

As at other times in its history, Trafalgar Square was part of this picture. There were not necessarily clear-cut boundaries among the various groups who used the Square; there was at least an ideological relationship, for example, between the union groups advocating better protection for working people and the Communist Party. Overall, the aim of political groups in using the Square was to make themselves visible, proclaim their policies, answer their critics and raise support and membership of their organisations. For example, in the 1920s, around 2,000 British Fascists rallied there, using it on Remembrance Day 1924 as a starting point for the short march down Whitehall to the Cenotaph to lay a wreath and to ensure that 'the Union Jack should hold sway on such a day and that the Red Flag should be banned'[45] because Remembrance Day coincided with the seventh anniversary of Bolshevist rule in Russia.[46] This remark by the Fascist organisers indicates their nationalist political agenda, but also suggests that they may have considered Trafalgar Square's imperial symbolism a suitable frame for their pro-British ideology.

---

44 Ward, *Britishness*, 104.
45 'British Fascists' Procession – Big Display', *The Manchester Guardian* (10 November 1924).
46 'British Fascisti – Trafalgar-Square Meeting', *The Times* (10 November 1924), 9.

On 15 October 1932, Oswald Mosley launched his British Union of Fascists with a rally in Trafalgar Square. The event was described as 'lively', and characterised by multiple interruptions from a hostile crowd.[47] By the late 1930s, the BUF had come to be associated by many with violence and almost paramilitary organisation. A big rally at the Olympia Stadium at Earl's Court in 1934 was marred by the very violent ejection of anti-fascist protesters in the crowd, in full view of the middle-class constituency it hoped to woo. This included the owner of the *Daily Mail*, Lord Rothermere, who had used his paper to support the group until then. In 1936 the 'Battle of Cable Street' in London's East End disintegrated into street clashes and police arrests, reinforcing reputation of the Blackshirts for a heavy reliance on the 'good old English fist' (sometimes stiffened with knuckledusters or lengths of rubber hosing). These disturbances were the background to a 4 July 1937 march by the BUF to Trafalgar Square. The route of the procession was planned through parts of the East End, effectively an act of occupation and assertion of power that would almost certainly have led to violence, as it had the previous year. This route was prohibited by the Home Secretary, and the BUF changed their plans. However, the marchers were still heckled by their opponents along the way, skirmishes in the Square led to twenty-seven arrests, and Mosley was reportedly barely audible over the jeers of the anti-fascists who ringed the Square – one unsympathetic report summed it up dismissively as 'some scuffles and a lot of noise'.[48] In the next sitting of parliament, the Home Secretary, Samuel Hoare, informed the House that 2,383 police officers had been deployed on 4 July, and police records estimated that 3,500 BUF supporters had participated (the number of their opponents was not discussed). The police contribution, it was argued by a Labour colleague, had effectively allowed the march to take place, and the BUF regarded the event as a success because

---

47   'Lively meeting in Trafalgar Square – Hostility towards Sir Oswald Mosley', *The Manchester Guardian* (16 October 1932), 21.

48   David Stephen Lewis, *Illusions of Grandeur: Mosley, Fascism and British Society, 1931–1981* (Manchester: Manchester University Press, 1987), 127 and 'Fascist March in London – 27 Arrests – Some Scuffles and a lot of Noise', *The Manchester Guardian* (5 July 1937), 9.

their opponents had not been able to prevent or curtail it, even if they had rendered the speeches unintelligible and engaged in numerous 'scuffles'.[49]

Given the pugilistic reputation of the BUF, the robust reaction of its opponents, and the numbers of participants who were ex-soldiers with first-hand experience of violence, the 'scuffles' were not surprising. Conflict between members of the crowd often occurred, with shouting over the speakers, 'jostling and pushing', fighting and rushing the platform reasonably common. Physical (and aural) dominance of the space was as material as it was symbolic on these occasions, and occupation of space appears to have constituted a successful protest. Such hostility was not unknown at other political rallies in the Square, including robust argument from the crowd, and sometimes, especially before the late 1920s when loudspeakers could be used for amplification, attempts to drown out the speaker. Lawrence has argued that that this tapped into longstanding traditions of political heckling based on the 'public's right to not only to interrogate, but also to torment, their would-be political masters' on the campaign trail, and that the 'rowdyism' that Mosley supporters engaged in (as had their detractors) had also been used by mainstream political parties to try and stop disruption to their meetings.[50]

The police were ever-present at events in the Square, and had the uncertain role of using violence to control the crowd. There are many descriptions of police behaviour on duty in the Square, from indulgent or good-natured during royal events or national celebrations to brutal and heavy-handed during demonstrations. During the 1920s and 30s, the constabulary were sometimes praised for rescuing people from serious mistreatment by other members of the crowd, but were also associated with the authorities that protests were aimed at, and so were singled out for criticism. A speaker at an October 1924 Communist demonstration, for example, exhorted protesters to resist this authority 'Why don't you revolt? Are you afraid of

---

49   House of Commons Debate 15 July 1937 vol 326 col 455–457; and Lewis, *Illusions*, 127.
50   Jon Lawrence, *Electing Our Masters: The Hustings in British Politics from Hogarth to Blair* (Oxford: OUP, 2009), 127.

Scotland Yard?'[51] This comment reveals something important about the version of Britishness in Trafalgar Square. The speaker, and many others like him, had an ambivalent relationship with the nation represented in the Square; they resented and often resisted the authorities that controlled the space, while appealing to a version of 'free-speech' Britishness that they accessed through the history of the site for counter-hegemonic protest. In encouraging listeners to revolt, the speaker identified the Square's role as an official site of control, its materiality shaped by and framed with imperial representations, and an irresistible location of resistance and occupation, the perfect place to start a 'revolt'. This is not to say that it was the only important site of political resistance or action. In London alone there were numerous halls and meeting places, small and large, outdoor spaces ranging in size and impact from Hyde Park to the streets of the East End. However, the Square has always retained special significance related to its location, history, use, prominence and symbolism that many groups have sought to activate in their particular claims and contributions to a British narrative.

The use of the Square by left and right-wing political groups in this period represented continuity with its political use in the past, in that organisations such as the Communist Party of Great Britain and the BUF were not mainstream political groups (although they may have enjoyed some mainstream support at times, such as Lord Rothermere's *Daily Mail*). As minority groups, the Square provided a valuable platform to raise their issues and press for policy change within the national government. Ward has suggested that the interwar period saw a 'domestication' of Britishness with 'the quiet, suburban and rural versions of national identity [...] in the ascendant', and with the aggression ascribed to fascists and then national socialist parties in Europe, particularly in the 1930s, constructed as a foreign contrast to peace-loving Britain.[52] To some degree, the use of the Square in this period supports this view, in that many of the demonstrations, rallies and protest meetings were related to workers' rights or state support

51   'Trafalgar Square Scenes – Attempt to Rush Communist Platform', *The Manchester Guardian* (13 October 1924), 9.
52   Ward, *Britishness*, 47.

for the unemployed and the disabled. On the other hand, the Square also framed British responses to international politics, as evidenced in its use by both Communist and Fascist organisations. Furthermore, as had always been the case, the imperial was still on display, primarily activated through major royal events.

Overall, by the time war with Germany was declared on 3 September 1939, the tone of the use of the Square was one of internal diversity within Britain. This diversity was pushed to the background somewhat during the war as national unity was actively promoted by officials for the purposes of production, security and public morale; for example, as early as April 1939, the Square was being used to recruit for civil defence.[53] At the end of the war in 1945, Trafalgar Square was a stage on which the official vision of Britishness was once more on display – although the narrative expressed within it was less likely to be of imperial reach and power and more that of a smaller and more vulnerable national community.

The 'People's War'

The version of national identity officially promoted in Britain throughout the Second World War was one of a national unity and common purpose. As Churchill said in the summer of 1940, after he had become Prime Minister: 'The whole of warring nations are engaged, not only soldiers, but the entire population, men, women and children. The fronts are everywhere'.[54] In wartime publications and broadcasts, writers and essayists such as George Orwell and J.B. Priestley reinforced the idea of a unified nation with long-standing universal characteristics, as in this example from Orwell's *England Your England*, first published in 1941:

53   'Mobile Recruiting Offices', *The Times* (25 April 1939), 20.
54   Angus Calder, *The People's War: Britain 1939–1945* (London: Pimlico, 2008 [1969]), 17.

Yes, there *is* something distinctive and recognizable in English civilization [...]. It is somehow bound up with solid breakfasts and gloomy Sundays, smoky towns and winding roads, green fields and red pillar-boxes. It has a flavour of its own. Moreover, it is continuous, it stretches into the future and the past, there is something in it that persists [...][55]

According to Rose, such themes of unity and endurance were part of a common narrative during the war that presented the British to themselves as a national community despite social divides such as class, region or gender:

> [During World War Two] There were numerous portraits depicting the nation as a unified community of ordinary people contributing to the war effort. These characterizations made 'the common man' central to the nation at war, celebrated diversity, implicitly advocated tolerance, and recognized Britain as a class – and gender-divided society but denied that it mattered to national unity – to the image of the British as essentially one people. This vision of World War Two patriotism provided the parameters for defining the nation as a community.[56]

In 1940, the phrase 'a People's war' was shorthand for an influential idea that 'shaped the rhetoric of five years of official and unofficial propaganda', and that actively promoted this idea of a unified nation.[57] The 'People's war' saw the civilian population effectively 'mobilised' in support of the fighting troops through schemes such as the Land Army and the conscription of factory workers. To reinforce further the notion of national unity, the government monitored civilian morale closely and called on British people to remain cheerful in the face of wartime deprivation such as increasingly stringent rationing. In the official wartime narrative, high morale was depicted as crucial to military success. This was reflected in Trafalgar Square by the use of Nelson's Column to erect hoardings calling for the public to 'Buy National War Bonds'.[58]

---

55   George Orwell, 'England Your England' in *Inside the Whale and Other Essays* (London: Penguin Books, 1957), 64.
56   Rose, *Which Peoples' War?*, 6–7.
57   Calder, *The People's War*, 138.
58   Hargreaves, *Trafalgar Square*, 54.

However, the official vision of a united and determined nation belied a much more complicated reality, and Britain at the end of the war was still divided by class, despite the attempts by the British government to unify the population.[59] By 1945 the country was exhausted, with a sense of vulnerability that extended to the 'national unity' that the authorities had worked hard to inculcate; Rose highlights the 'fragility of a unitary national identity, even during a war that involved total mobilization of the country's citizenry'.[60] In 1945, the possibility of a single British national identity was perceived as both stronger and weaker than it had been before 1939, as the wartime experience had both unified the nation towards a common purpose and exacerbated existing social divisions because of the difference in resources that people were able to draw on. MP and diarist Harold Nicholson, for example, recorded rumours that the 'King and Queen had been booed the other day when they visited the destroyed areas'.[61] When Buckingham Palace was bombed on 13 September 1940, the Queen was reported to have said that she was glad the royal residence had been damaged, because now she could 'look the East End in the face',[62] an area that was bombed extensively during the September 1940–May 1941 Blitz. Thus, despite official attempts to promote national unity in the face of physical and emotional wartime loss and trauma, the image of Britain as unified against its enemies was not the only experience of many Londoners, and the national community was not as uniform as the phrase 'the People's war' suggested.[63]

Another aspect of attempts to build a unifying national narrative was geographical. Although many areas of the UK were damaged by war,

59 See Rose, *Which Peoples' War?* and Calder, *The People's War.*
60 Rose, *Which Peoples' War?*, 2.
61 Calder, *The People's War*, 164.
62 *Ibid.*, 168.
63 See Rose, *Which People's War?*; Calder, *The People's War*; Gareth Stedman Jones, 'The "cockney" and the nation, 1780–1988' in David Feldman, and Gareth Stedman Jones, eds, *Metropolis London: Histories and Representations since 1800* (London: Routledge, 1989); and Mark Connelly, *We Can Take It! Britain and the Memory of the Second World War* (London: Longman, 2004).

attacks on London came to symbolise attacks on the whole country. The
Blitz strengthened the role of London as a national symbol, both in terms
of its built environment and its inhabitants. London residents were cast
as plucky, stoic and resilient, and during the Blitz the term 'Londoner'
became a 'title of honour': 'from 7 September 1940, literally overnight,
the image of the Londoner was remade in the eyes of the English speaking
world [... as the war] bridged the gulf between London and the nation'.[64] If
the nation looked to Londoners as symbolic of stoic endurance, London's
built environment also became a symbol of British resilience, and played
an important role in the narrative of national resistance to German aggression, especially after the fall of France in 1940.[65] In particular, the Blitz of
September 1940 to May 1941 brought London to the forefront of the British
national imagination, placing it 'at the heart of the nation'.[66] As evidence of
London's power to symbolise national survival, at the end of the war in a
speech on 9 May 1945, Churchill described it with imperial and anthropo-
morphic overtones as a robust and resilient African beast: 'London, like a
great rhinoceros, a great hippopotamus, saying: "Let them do their worst.
London can take it."'[67] An example of Trafalgar Square's centrality in this
landscape was evident in *From the Four Corners*, a Ministry of Information
film that saw soldiers from Canada, New Zealand and Australia converge
on the Square in search of a pint. Rescued from a gushing middle-class
woman who asks 'Are you boys from the *Empire*?', the men are treated to
a heartfelt soliloquy by Leslie Howard on London as a symbolic landscape
of 'justice, tolerance and the rights of man'.[68]

According to Bell, London's built environment represented national
resilience, because although it was a military target, it was also a 'powerful

---

64   White, *London in the 20th Century*, 102.

65   Amy Helen Bell, 'Landscapes of Fear: Wartime London, 1939–1945', *Journal of British
       Studies* 48 (2009), 153–157.

66   Gareth Stedman Jones, 'The 'cockney'', 314.

67   Charles Eade, ed., *Victory: War Speeches by the Right Hon. Winston S. Churchill,
       O.M., C.H., M.P.* (Melbourne: Cassell and Company Ltd, 1945), 127.

68   *From the Four Corners*, Denham and Pinewood Studios, 1941. See also Wendy Webster,
       Englishness and Empire 1939–1965 (Oxford: OUP, 2005), 29–30.

ideological symbol of civilian endurance'.[69] The importance of the built environment was based in large part on its cultural symbolism. Well-known images of the Blitz, such as St Paul's Cathedral intact despite German bombardment of the surrounding area, were (and still are) part of the narrative of British endurance as told through London itself. A bomb fell just opposite Trafalgar Square in 1942, but a post-war Ministry of Works report continued the tone of resilience, reporting that 'in spite of bombs falling on every side of Trafalgar Square the column and the statue [...] suffered little'.[70] Throughout the war, hoardings erected around the base of Nelson's Column exhorted passers-by to 'Buy Defence Bonds' and loudspeakers were installed in the fountains to play music from a military band.[71]

However, the fragility of the urban landscape was evident. London towards the end of the war was uninspiring, with 'buildings black with soot [...]. Boarded-up windows [...] mean terraces, with now and then a gap [...] acres of bomb sites'.[72] Fear had been a feature of the war for many Londoners, and with good reason: in the 1940–1941 bombardment, 15,775 Londoners were killed and almost one-sixth made homeless.[73] London's built environment, as well as its population, was a casualty of the Blitz, and according to Porter, this had a significant effect on the city: 'Killing only 20,000 civilians, the Blitz destroyed or damaged 3.5 million homes in metropolitan London. London's wartime toll was thus less in deaths or social breakdown than in destruction to property'.[74] This included a bomb that fell just to the south of Nelson's Column in 1940. It entered the escalator shaft in Trafalgar Square tube station and killed seven people and injured thirty-three who were sheltering there.[75]

69   Bell, 'Landscape of Fear', 158.
70   'Our London Correspondence – Admiral's Inspection', *The Manchester Guardian* (6 March 1946), 4.
71   See Hargreaves, *Trafalgar Square*, 54 and 'Loudspeakers in Trafalgar Square, London, England, UK, 1941, in the Imperial War Museum's collection, < http://www.iwm. org.uk/collections/item/object/205195856 > accessed 10 April 2013.
72   Waller, *London 1945*, 1.
73   Bell, 'Landscape of Fear', 158.
74   Porter, *London*, 341.
75   Hargreaves, *Trafalgar Square*, 55.

Although the Square was not itself badly damaged, rocket attacks transformed the city into a 'landscape of fear':

> During the Second World War, London was the primary and most dramatic stage where both personal and communal fears were played out. Londoners feared death, loss of property, injury, bereavement, and on a broader scale, the loss of the war and the end of Britain [...] the landscape of wartime London, stripped of familiar landmarks and after 1940 damaged by bombs [...] became imbued with fears of imminent individual and collective destruction.[76]

Overall, during World War Two, London's people and built environment represented contradictory aspects of British society and identity. The social unity that government morale-boosting campaigns sought to engender appeared to paper over existing class divisions.[77] The survival of buildings such as St Paul's Cathedral, the heavily bombed East End and the image of the 'plucky cockney' were symbolic of national survival and resilience, and the buildings and place names of the central areas invoked an enduring global Empire. On the other hand, the destruction of property and the displacement of thousands of Londoners symbolised national fragility. Because of this special and complex role in the national imagination, central London's VE Day celebrations had a national significance beyond those in other locations. During the celebrations, Trafalgar Square was a popular destination for revellers and it helped to shape how the nation was experienced and understood.

76   Bell, 'Landscape of Fear', 153–157.
77   Rose, *Which People's War?*

VE Day in Trafalgar Square

On 7 May 1945, North Americans and Europeans, including people in Britain who had access to German broadcasts, heard that Germany had surrendered and that the war in Europe had ended. In the UK, the BBC briefly covered German reports of surrender in its 3.00 pm afternoon bulletin. However, despite widespread rumours of the end of hostilities, there was no official announcement in Britain about the end of the war during the afternoon, and the lack of official communication created a sense of frustrated expectation amongst the public.[78] This delay confused and annoyed many people. According to Mass Observation (MO) observers, as the day passed, the general public became cynical and suspicious of government motives, and 'a note of bitterness crept in, and a tendency to blame the government for what they vaguely felt was a deliberate policy of procrastination'.[79]

The impending end of the war, and the arrangements for the public announcement, had been discussed in a War Cabinet meeting around two weeks before, on 25 May.[80] The delay in telling the public that the war had ended was partially a result of an existing agreement with Stalin that the official announcement would be made simultaneously in the UK, the US and Russia. Despite growing public knowledge of the German surrender, Stalin did not want to deviate from this agreement and the War Cabinet 'thought it preferable to avoid the risk of a reproach from Marshal Stalin'.[81] Furthermore, Cabinet documents show that officials anticipated mass absenteeism as a result of the end of the war and that they questioned the necessity for an official 'VE Day' at all, based on concerns over disruption to work and 'amongst civilians a relaxation of spirit, which may have an adverse

---

78   Russell Miller, *Ten Days in May: The People's Story of VE Day* (London: Michael Joseph, 2007 [1995]).
79   Mass Observation Archive FR 2263 1945, 7.
80   CAB/65/50/22, 342.
81   *Ibid.*

effect on the conduct of the war in the East'.[82] The 9.00 pm announcement on the evening of 7 May 1945, therefore, represented a compromise between the demands of British allies and an official desire to manage the public reaction to the war's end. In addition to foreshadowing the King's speech scheduled for the next afternoon, the broadcast also announced that 8 and 9 May had been designated as public holidays. In Trafalgar Square, where people had been waiting for the evening news to be broadcast on public speakers, the reaction to this announcement was subdued:

> People hanging around in knots waiting for the nine o'clock news. Seats round the fountain difficult to find. Nine o'clock and the news is on [the radio, over loudspeakers]. For a few seconds there's a strange stillness in the square as people listen to what the announcer is saying. This time the waiting crowd hear that Mr Churchill is to speak at 3 o'clock tomorrow and that Wednesday is to be VE Day [...]. Quietly and soberly the crowd disperses.[83]

This example shows that, rather than being met with spontaneous celebration, the initial announcement of the end of the war was low-key and was met with a measure of public cynicism, frustration, or apathy. According to these accounts, the initial British response was slow to build; in part because of the staggered nature of the announcement, the initial public reaction was not one of wild, spontaneous rejoicing. The popular picture of joyous national celebration was moderated by the ambivalence that many people seemed to feel, and the sense that there was not much to celebrate beyond the enjoyment of a couple of days off work. Thousands had lost friends or relatives, been injured themselves, or had been subjected to other personal trauma, such as the destruction of property. Daily life was still dominated by shortages and rationing, and the war against Japan was ongoing.

Later in the evening on 7 May, however, the public mood had become more lively. The West End, including Trafalgar Square, refilled with people, as described by a woman who lived nearby:

> The whole square was filled with people. One could just see groups of men and women, their arms linked together, whirling round and round. Others leapt about

---

82 CAB 66/65/19 1945, 1.
83 Mass Observation Archive FR 2263 1945, 24.

on their own in their irrepressible relief and joy [...]. An enormous tide, or river, of humanity filled the square in ever-increasing numbers, as others heard the news and flocked in from neighbouring streets. The great lions, occasionally visible in the flare of a torch, seemed the right background to the spontaneous expression of relief of those thousands who, for so long, had endured the shattering sorrows of war, the darkness and gloom contrasting with the light and joy in the hearts of the people.[84]

This passage above identifies two important themes. First, the experience of using the streets of central London to celebrate with thousands of others: 'an enormous tide, or river, of humanity' was a common type of description in individual accounts of the celebrations. This created a sense of crowd-feeling that enhanced the occasion, contributing to the excitement felt by participants and linking them to others. Second, particular buildings and monuments, including Landseer's lions revealed in 'the flare of a torch' or Nelson's Column, which on the night of 8 May was floodlit, were transformed into metaphors for national survival and resilience. Thus, the official mobilisation of the physical national symbols in the Square's built environment helped shape popular responses to the celebrations.

Crowds and 'group-feeling'

It was fine and warm on 8 May, and thousands of people thronged in the streets of central London, where authorities planned for large numbers of people to gather. In particular, the police foresaw large crowds in Trafalgar Square, and planned for the impact on traffic should the Square overfill and crowds block adjoining streets. The interdepartmental conference on VE Day arrangements was clear about the impact on London of the cease-fire:

> Notwithstanding any suggestion the Government may make that celebrations should take place as locally as possibly, crowds will undoubtedly throng in the centre of London and will congregate round such places as Buckingham Palace and Trafalgar Square.[85]

---

84   Miller, *Ten Days*, 131.
85   HO 186/2050, 3.

In several accounts of VE Day, the crowds of other people celebrating the same event was an important part of the spectacle of the event, and this was enhanced by the more general colour and excitement of the occasion. The MO's editors summed up the overall mood:

> Gradually the flags went up until London was 'vivid with red, white and blue'. Flags and decorations and the excitement and group feeling which is associated with them, must have helped a lot to lift people from the mood of apathy and depression into which they sunk in the last week of the European war.[86]

Here, the visible, colourful decorations in London were a counterpoint to public ambivalence about the announcement of the end of the war, and helped to create a feeling of community among celebrants that drew people into the festivities. In addition to the decorations, descriptions of the size and behaviour of the crowd were common aspects of MO accounts of VE Day in central London, and the presence of 'group feeling' identified above is reinforced in subsequent reports. The spectacle of the crowd appears to have been a large part of the appeal of the celebrations, as was the experience of moving through the crowded central London landscape.

Overall, two aspects of the crowds' behaviour emerge from these reports from VE Day. The first was of calmness or even resignation in the face of a long wait for an announcement or the appearance of Churchill or the Royal family. In general, there was little reported conflict despite the large numbers of people packed into a relatively small area. In central London, for example, 'the police were massively outnumbered but there were few reported cases of trouble. Most people were heading in similar directions – towards Buckingham Palace, Trafalgar Square, the Houses of Parliament and Piccadilly'.[87] MO investigators reinforce the characterisation of the crowds as relatively quiet and peaceful, although this appeared in part to be due to public confusion over the correct response to the occasion: 'everywhere, people are sitting on walls or walking around with a curious aimlessness. It all seems very muddled and confused for a

---

86   Mass Observation Archive FR 2263 1945, 15.
87   Craig Cabell and Allan Richards, *VE Day: A Day to Remember* (Long Preston: Magna Large Print Books, 2005), 132–133.

day of celebration'.[88] In contrast, the other main narrative concerning the celebrations, most common in the popular press, was of a city going 'wild' with elation at the end of the European war. As the evening wore on and, presumably, as the pubs carried on a brisk trade, the crowds became more lively. The *Daily Mirror* printed photographs of people dancing in the streets and climbing on lampposts in Piccadilly Circus on the evening of 7 May under the banner headline 'The Nation Celebrates'. The back page of the paper featured a large picture of a crowd in Piccadilly Circus, with an accompanying caption that described a feeling of elation, although it seemed to be as much about the prospect of a holiday as the end of the war in Europe: 'With the prospect of two whole days' holiday in front of them, people let their pent-up feelings go'.[89] Similarly, the MO explained public enthusiasm in terms related to the embodied experience of being in the crowd that generated public excitement:

> [...] the feeling of being in a crowd, the decorations and festive atmosphere, perhaps the final, official certainty of peace [...] managed at last to stir people to something like real excitement [...]. Most people's participation was confined to singing and cheering when called upon, and admiring the abandonment of others.[90]

These accounts point to the role of the crowd in creating a sense of unity and inclusive national celebration, a contrast to the more general social division that Rose and Calder identify as being evident in wartime. Here, the crowd itself appeared to represent the nation and the feeling of being a member of the crowd engendered excitement and a sense of celebration. In this unofficial discourse, the nation was imagined as the crowd on the streets of central London. Furthermore, media coverage of the event highlighted this aspect. For example, the crowd represented 'the spontaneous rejoicing of a whole nation',[91] because of participants coming from 'every part of the land'.[92]

88   Mass Observation Archive FR 2263 1945, 36.
89   'VE-Scene Trafalgar Square', *The Daily Mirror* (8 May 1945), 8.
90   Mass Observation Archive FR 2263 1945, 42.
91   'A people's heart', *The Daily Express* (9 May 1945), 2.
92   'ATS girls light up St Paul's, Londoners "go to town"', *The Daily Mail* (9 May 1945), 3.

Furthermore, reporting on the event framed the activities of the crowd with references to central London's landmarks, including Trafalgar Square. The experience of being in the crowd appeared to be conditioned by the built environment, and the national and imperial landmarks of Trafalgar Square, Whitehall and the Houses of Parliament, and Buckingham Palace, all provided a symbolic setting for the celebrations. VE Day was presented in the media as a public performance of community in the historically resonant built environment of central London, one that included 'ceremonial avenues and spaces designed for spectacular performances'.[93]

The use of these places, including the journey from site to site alongside others in the crowd, was part of the experience of how national identity was understood on 8 May 1945. Records of the occasion described the movement of people from place to place, especially in the crowded areas within central London and the West End, particularly around the area bounded by Trafalgar Square, Leicester Square, Piccadilly Circus, Buckingham Palace and the Houses of Parliament. Trafalgar Square was one site within a larger area that people invested with meaning during the VE Day celebrations, as in this example from the *Daily Mail*:

> To The Empire's Heart Came the greatest Victory Crowd
>
> In Trafalgar-square, symbolic heart of the British Empire, there assembled yesterday the greatest victory crowd in the history of a nation. They came, these many thousands, not only from all parts of London, but from every part of the land that could afford them access to the Mother-Capital. Many hundreds of them must have walked miles [...] but they kept their Rendezvous with Triumph.[94]

Meanwhile, in the *Daily Mail* of the evening of 7 May, the Square provided a stage for congregation amidst the symbols of Empire:

> It was a high old time in Trafalgar Square last night. Everybody wanted to climb something. [A] party of Wrens and Allied soldiers celebrated by clambering on to the lion. Army police men present – like Nelson on his column – turned a blind eye.[95]

93   Driver and Gilbert, *Heart of Empire?*, 10.
94   'ATS Girls', *The Daily Mail*, 3.
95   Quoted in Cabell and Richards, *VE Day*, 27.

In these examples, the Square was a 'reminder of nationhood [that] serve[d] to turn background space into homeland space'.[96] Many media reports of VE Day highlighted the connection between a built environment charged with national and imperial symbols, and the experience of celebrating a national event alongside thousands of other people; even if revellers themselves did not stop and pause to consider the imperial landscape in which they celebrated, it still framed their experiences.

The crowd was also the audience for its own spectacle, and individuals seem to have observed as much as participated, with part of the significance of the day residing in the effect of being with large numbers of like-minded people. In Anderson's terms, these examples demonstrate how the imagined community of the nation was made real during these celebrations, as individuals included themselves in a larger group celebrating the same event.[97] They also support Rose's claim that an individual's understanding of national identity is based on their experience of 'taking action' with others, that 'community may be understood to be the outcome of the process of collective identification; it is made through practices that establish who "we" are as a collective body'.[98] In the case of VE Day, the use of central London by the crowds was an example of 'people [taking] action together' in a way that helped them to imagine themselves as one community. The existing social divisions within London in 1945 appear to have been largely forgotten in the moment of national celebration in Trafalgar Square and other areas of central London. First-hand and media accounts do not focus on class or other categories of social division, instead emphasising the unity of purpose of the VE Day crowd.

However, the version of the use of Trafalgar Square and surrounding areas during VE Day celebrations, drawn from media and personal accounts above, does not tell the whole story of the occasion. For example, Kynaston

---

96 Billig, *Banal Nationalism*, 43. The reference to Allied soldiers and to the 'Empire's Heart' also taps into a narrative of what Wendy Webster calls 'the People's Empire', in which imperial institutions are depicted as constituted by and for the individuals they regulate – see Webster, *Englishness and Empire*, Chapter 2.

97 See Anderson, *Imagined Communities*.

98 Rose, *Which People's War?*, 10.

points out that 'The West End [...] was not London, let alone Britain'. He cites the MO's overall assessment of the evening:

> Usually, crowds were too few and too thin to inspire much feeling [...] and on V.E. night most people were either at home, at small private parties, at indoor dances or in public houses, or collected in small groups around the bonfires, where there was sometimes singing and dancing, but by no means riotously.[99]

London's role as the national capital meant that it dominated reporting of the event, but other, less spectacular experiences were more characteristic of the experience of VE Day. Outside of central London, and even for many people within it, the occasion was not a wild party on the streets, or a chance to engage with a much larger national group, but was instead a quiet event spent alone or with a few friends or neighbours. As multiple narratives were constructed about the event in public celebrations in Trafalgar Square and central London, other, more private stories also wove together the national and the personal in many different ways.

'Dazzling relief'

As people moved through central London, stopping at Trafalgar Square, Buckingham Palace or along Whitehall, they observed more than just their fellow celebrants. According to Calder, the illumination of central London was an important part of the festive atmosphere of the evening of 8 May, as 'The night scene was consummated by the exhilaration of full-powered, aggressively sported, lighting. The statues and public buildings were floodlit; the searchlights danced a ballet in the sky [...] [there was a] glow which seemed unearthly to the small children whose parents let them stay up late to wander in this unimagined fairyland of illumination.'[100]

---

99 David Kynaston, *Austerity Britain 1945–51* (London: Bloomsbury, 2007), 13.
100 Calder, *The People's War*, 567.

A reaction to the display in Trafalgar Square, Mary Carlton's first-hand account captures this sense that the spectacle of the colourful lights was a source of wonder:

> [Trafalgar Square] was really a picture. Green floodlighting on Nelson, Mauve on the lions, Green and White on Admiralty Arch, White on the Art Gallery and St Martin's and there was a large globe of changing lights on top of the Coliseum. It was wonderful.[101]

The display had both practical and symbolic significance. As well as lighting the streets for the thousands of people in the city, it was as if 'through representations and rhetoric [the nation] appear[ed] to exist in a concrete form'.[102] An example is the following excerpt from the *Illustrated London News*, which contrasts the brilliance of the lights with the metaphorical and literal darkness of the war and the blackout:

> As at the outbreak of war one of the most poignant phrases was 'The lights are going out all over Europe', so it was most fitting that the lights of London should blaze out to celebrate Victory in Europe: and Victory Night was celebrated with lights of every form, colour and source [...] that peculiarly modern art of peace – floodlighting – was used to illuminate and to illustrate all the main buildings of the metropolis [...] some of the chief landmarks of London, glowing in the unaccustomed light [included] [...] Admiralty Arch, the Royal Exchange, Nelson's Column and the Houses of Parliament.[103]

Here, the illumination of the 'main buildings of the metropolis' represents national victory and a metaphorical return to the light of peace. Along with other important structures, 'searchlights were symbolically trained onto London's most iconic landmark, Nelson's Column' on the evening

---

101 From Mary Carlton's 'Memories of VE Day' on the BBC's project 'WW2 People's War', http://www.bbc.co.uk/history/ww2peopleswar/stories/29/a4143629.shtml> accessed 10 April 2013.

102 Rose, *Which People's War?*, 7.

103 'The lights of London: Famous Buildings Floodlit on Victory Night', *The Illustrated London News* (19 May 1945), 523. This is a reference to what Foreign Secretary Edward Grey said at the outbreak of the First World War: 'The lamps are going out all over Europe'.

of 8 May.[104] The floodlighting was symbolic in terms of national survival, especially in the context of war damage and the blackout:

> as dusk fell the floodlights, doubly astonishing after nearly six years of darkness, were switched on, throwing into dazzling relief all those great public buildings which had miraculously survived [...] most impressive of all [...] was the stately quadrangle of buildings around Trafalgar Square, with [...] 'Horatio Nelson standing aloft on a greenish ray of light [...] as romantic as even he could have wished'.[105]

This aspect of the celebrations was widely reported in local and national newspapers. According to the *Daily Telegraph*, 'Enormous crowds assembled in Westminster, Whitehall and Trafalgar Square last night to see some of London's most historic buildings floodlit for the first time since the [1937] Coronation'.[106] Here, Trafalgar Square evoked previous celebrations of a powerful imperial event, the coronation of King George VI. Floodlighting Nelson's Column helped to connect the present to the national past, and acted as a reminder of pre-war history and the survival of the monument despite the bombings that London endured during the war, including a hit very close to the Column itself. The references to the brilliant, blazing glare of floodlights used the built environment to create a narrative that invoked national survival, a demonstration that the buildings, like the nation, had endured years of war. The reporting in *The Daily Mail* also alludes to the historical symbolism of the Square, linking this to the behaviour of the crowd, describing it almost as a single entity:

> A floodlit London went crazy with joy last night [...]. At 10 p.m. Trafalgar-square was flooded with light as Nelson's Column was illuminated, and the swaying crowd of 100,000 had hardly room to breathe. Searchlights placed in front of the National Gallery lit up the historic scene and their beams reached along Whitehall [...]. The crowd shrieked and roared when coloured rockets were sent shooting into the air. Some cast a fountain of light almost as high as the top of Nelson's Column.[107]

104 Hargreaves, *Trafalgar Square*, 8.
105 Longmate, *When We Won*, 72–73.
106 'London Revels in Glare of Floodlights', *The Daily Telegraph and Morning Post* (9 May 1945), 5.
107 'ATS Girls', *Daily Mail*.

Even though the *The Daily Mail's* main emphasis is the crowd, it still describes the Square, with Nelson's column and the façade of the National Gallery, as 'historic', contextualising the celebrations there with the representation of the national past that it identified as significant. Through reference to the Square and its structures, the report links a giddy, celebratory present with a narrative of national history.

For the authorities, the landscape of central London, including Trafalgar Square, framed the VE day celebrations with a range of spatial reminders of the nation, constructing VE Day as an event 'situated within a homeland, which itself is situated within a world of nations'.[108] Authorities' arrangements to spotlight nationally significant structures such as Nelson's Column reminded participants and observers of an official version of British history, and the event emphasised the symbolic value of the built environment as redolent of previous British victories and power. By spotlighting major London landmarks, the authorities emphasised an urban landscape punctuated by national and imperial symbols that both reminded viewers of the nation and connected them to a version of the national past. At the same time, it reminded participants of wartime loss and destruction, as well as their survival of the war. Importantly, in media coverage of the celebrations London's damaged landscape was central to the narrative of British resilience during the war, and this landscape cast victory in terms of survival rather than dominance.

Although the 'occupation' of the Square during the VE Day celebrations was not adversarial or contested in the same sense as the Suffragettes' use of the space, it did represent an officially sanctioned version of the nation that the media promulgated and which featured 'the People' as both its subject and its audience. Nelson's Column symbolised British resilience in the teeth of German bombardment, and its survival was visually emphasised alongside other floodlit landmarks, including the National Gallery, Admiralty Arch, Buckingham Palace and St Paul's Cathedral. The choice of these buildings and monuments reflected an official narrative of British history and identity and linked the end of the European war with

108 Billig, *Banal Nationalism*, 8.

other important aspects of Britain's official culture, such as naval power, the monarchy and the church. However, such 'official' narratives were complicated by first-hand accounts, which stressed the embodied response to 'being in a crowd' and the spectacle of the event. While first-hand reports appear to imagine the nation as the crowds in the street of London, it was the emotional rather than the physical urban landscape that loomed largest.

In 1945, despite authorities' efforts to represent a homogeneous picture of national identity through the use of phrases such as 'the People's war' or symbolic floodlighting of national buildings, Trafalgar Square provided the means for a subtle reimagining of this official image of Britain as grounded in a particular historical narrative. By focusing on the activities of the crowd, 'group-feeling' and the spectacle rather than the explicit symbolism of the floodlighting, individual participants do not appear to have been as concerned with the national symbolic significance of these events as authorities would have liked, or even as their behaviour indicated, as represented in the press. Trafalgar Square provided the possibility for gentle resistance to the powerful official national narrative anchored in the past, because participants' accounts appeared to be very much focused on the present. Similarly, in the interwar period, the Square was used to express domestic concerns or to press for greater visibility in British rather than imperial politics. However, this gradual change of focus to national issues, alongside the occasional spectacular, imperial event, was complicated in the post-war period as concerns about national rebuilding were accompanied by decolonisation and the anxieties about the Cold War, as well as important changes to London's racial and cultural makeup.

1 Statue of General Sir Henry Havelock in the Square's southeast corner.
Photo: author's own.

CHARLES JAMES NAPIER

GENERAL

BORN

MDCCLXXXII

DIED

MDCCCLIII

ERECTED BY PUBLIC SUBSCRIPTION

THE MOST NUMEROUS CONTRIBUTORS

2 Statue of General Sir Charles Napier in the Square's southwest corner.
Photo: author's own.

3 Emmeline Pankhurst invites the crowd in Trafalgar Square to 'rush the House of Commons', 1908. Photo: Getty Images.

4 Women's Auxiliary Army Corps recruits new members during the First World War.
Photo: Imperial War Museum.

5 Trafalgar Square bombed, 1940. Photo: Keystone/Getty Images.

6   A sailor and two companions tidy up after a night spent in Trafalgar Square, 1945.
Photo: Mirrorpix.

7 Spectators climb one of the Square's lions during the funeral of George VI, 1952.
Photo: Henri Cartier-Bresson/Magnum.

8 Protestors re-enact the Sharpeville massacre in Trafalgar Square, 1970.
Photo: Roger Jackson/Getty Images.

CHAPTER 5

The Cold War and the Colonies (1948–1990)

The domestic focus present during the VE Day celebrations – evident in the floodlighting of important buildings and the local crowd-feeling that was reported by participants – intensified for many people in the decade after the war. Reconstruction of British cities and industry occured slowly, and into the 1950s the country was muffled by a sense of pessimism. Christopher Isherwood, visiting London in winter 1947, was confronted and depressed by London's 'physical shabbiness': 'Plaster was peeling from even the most fashionable squares and crescents; hardly a building was freshly painted […]. Many once stylish restaurants were now reduced to drabness and even squalor […]. London remembered the past and was ashamed of its present appearance. Several Londoners I talked to at that time believed it would never recover. "This is a dying city," one of them told me'.[1] The 'landscape of fear' that described London during the war appeared to remain relevant in its traumatised post-war fabric.

In terms of its external relationships, the British Empire was changing to a Commonwealth of independent states, a process of decolonisation that was reflected in changing patterns of immigration to the UK. The relationship with Europe was transformed by the entry of the UK into the European Economic Community in 1973, ending the previous system of preferential trading with Commonwealth countries, while the 'special relationship' with the US continued to strengthen post-war. As London was slowly rebuilt, these changes to global politics and society were reflected in its landscape and use, and Trafalgar Square played a concomitant role in public debates about Britain's place in the world.

---

1   Kynaston, *Austerity Britain*, 191.

This chapter focuses on the relationship between the use of Trafalgar Square and ways of thinking about British national identity from the end of the Second World War to the end of the Cold War. Although this is a long period to cover in one chapter, this is made easier because first, the themes that made the Square important earlier in the twentieth century persist: it helps to construct Britishness through reference to its imperial history; by its public prominence and value to minority organisations seeking to raise publicity; and secondly, because historical uses of the Square were used to frame and legitimise later uses. Furthermore, throughout this period, the physical space of the Square continued to act as a symbolic substitute for the metaphorical space of the nation, with some protest groups using it to stake a claim to Britishness.

With the notable exception of the Poll Tax Riot in 1990, protest in the Square in this period did not appear to have a drastic effect on official policy, with governments apparently unruffled by even the largest rallies. Still, groups valued the ability to mount very large demonstrations for their value in showing the extent of support for protest causes to the general public, and the Square remained a prominent site. The Square was also important in showing protest groups to *themselves*: it helped to maintain support and helped minority interests retain a stake in the national picture. The use of the Square in this period continued to reveal the complexity of how and by whom Britishness was defined; in addition to national identity, gender, age, race, class or urban and imperial identities came into focus at various times. To draw out these arguments, this chapter focuses in most detail on the use of the Square by two groups that used it from the 1960s to the 1990s: the Anti-Apartheid Movement (AAM) and the Campaign for Nuclear Disarmament (CND).[2] These two groups form the central focus because they demonstrate the global focus that was evident in the Square during this period. The issues they campaigned about show how domestic social, economic and political transformations were linked to larger international issues. The AAM case study provides a window into questions of race relations, migration and Britain's place in a decolonising

---

2   Gavin Brown, a human geographer, is compiling material on the non-stop picket of
     South Africa House from 1986–1990. See <http://nonstopagainstapartheid.word-
     press.com/ > accessed 18 April 2013.

world, whereas the discussion of the CND shows how British identity was defined in light of the Cold War and the proliferation of nuclear arms.

The discussion will be framed within larger questions regarding decolonisation, beginning with a brief account of the arrival of West Indian immigrants on the *Empire Windrush* in 1948, which marked the beginning of contemporary 'multicultural Britain'. This milestone arrival was part of a much larger process that continued in the 1960s, the migration to Britain by people from the West Indies, Uganda and other British colonies. Although people of black and Asian backgrounds had been present in Britain and London for centuries,[3] the arrival of the *Windrush* brought a highly visible new population to London. Less than twenty years later, in the 1964 British General Election, a Conservative candidate in the West Midlands seat of Smethwick campaigned for, and won, a seat under the slogan 'If you want a nigger for a neighbour, vote Labour'. By 1968, Enoch Powell was predicting widespread social catastrophe as a result of the scale of immigration from former colonies, and claimed to reflect what 'thousands and hundreds of thousands [of white Britons] are saying and thinking' in demanding that 'Commonwealth immigration' be stopped and those immigrants already in the UK be encouraged to leave.[4] Then as now, immigration to Britain was a politically contentious issue.

Against this background of domestic racial politics, the British AAM took root and grew. From 1963 to 1990, this movement demanded change to the official South African political and social system of racial segregation and discrimination that reflected a new relationship with former colonies, and that developed alongside changing attitudes to race and immigration in Britain. They used a combination of tactics that included boycotts of South African products and sporting teams, public protests, meetings and demonstrations, and direct picketing of South Africa House, on the eastern edge of Trafalgar Square.

---

3   See Peter Fryer, *Staying Power: The History of Black People in Britain* (London: PlutoPress, 2010 [1984]); and Robert Winder, *Bloody Foreigners: The History of Immigration to Britain* (London: Abacus, 2004).

4   In 'Enoch Powell's "Rivers of Blood" Speech', *The Telegraph* (6 November 2007) <http://www.telegraph.co.uk/comment/3643823/Enoch-Powells-Rivers-of-Blood-speech.html> accessed 6 February 2013.

The timeline and content of the AAM's use of the Square overlaps with the activities of CND, an organisation whose agenda was framed by the global politics of the Cold War. Left-wing reaction to the Cold War was volubly expressed in the Square, with its well-known use by CND in the early sixties as a final rallying point for marchers from Aldermaston or High Wycombe, both US military installations in Britain. Both the AAM and the CND were domestic expressions of public debate about Britain's international relationships and the nature of its role in a decolonising and nuclear-armed world. By the end of the Cold War, much of this debate coalesced around Margaret Thatcher's leadership of the country; it is with this in mind that the chapter concludes with a discussion of the violent response to the Poll Tax in 1990 that saw her Prime Ministership emphatically undermined. This was the most violent event in the Square for many years, the scale and extent of which had not been seen for a century, and which reprised the same arguments about criminality, poverty and public expression that surrounded much earlier violent demonstrations.

First, however, this chapter briefly considers the Square directly after the war, when British reconstruction demanded an intensely domestic focus, and the Square was a site of both protest and spectacle. Royal events also continued to be important, ritual uses that linked the site to earlier versions of Britishness that were often received as deeply traditional, despite their relatively short histories.[5] The first section concludes with Winston Churchill's funeral, an event that has been described as marking the end of Empire, and which also embedded the Square in an official landscape of national display.[6]

5   See Cannadine, *The British Monarchy*.
6   See Webster, *Englishness and Empire*.

Public spectacle and imperial ritual

Britain was exhausted after the war, with its people subject to shortages and strict austerity measures: 'meat rationed, butter rationed, lard rationed, margarine rationed, sugar rationed, tea rationed, cheese rationed [...] make do and mend'.[7] In addition to food shortages, the country faced record debt, poor quality housing, and a massive rebuilding task. The public mood was made worse by coal shortages and the bitterly cold winter of 1946–1947, with freezing weather and coal shortages that led to rolling blackouts and very uncomfortable conditions.[8] People worked by candlelight during the day due to lack of electricity, and bundled on layers of clothing to keep warm indoors.[9] The disgruntled undercurrent occasionally broke out into protest, including against bread rationing. One such July 1946 demonstration of 'many housewives from all over Britain' marching 'through Whitehall to Trafalgar Square' was reported as far away as Tasmania.[10] The protesters, organised by the conservative British Housewives' League, presented a petition to the King after a rally in the Square at which they reportedly endured 'a great deal of interruption, apparently from youths'.[11]

As in the past, alongside its use for protest or demonstration, Trafalgar Square retained its capacity to dazzle, and the London crowds responded enthusiastically to opportunities for diversion. In 1948 the Square was partially refurbished to include lighted fountains, like 'jellies in ingenious bowls which [...] change colours before you have time to dig a spoon into them'.[12] The crowd's enjoyment of them was detailed in the *Manchester Guardian*:

7   Kynaston, *Austerity Britain*, 19.
8   German and Rees, *The People's History*, 230.
9   Kynaston, *Austerity Britain*, 185–205.
10   'Protest Against Bread Rationing', *The Advocate* (16 July 1946), 1.
11   'Bread Petitions to the King – Women's demonstration against rationing', *The Times* (15 July 1946), 4.
12   'Miscellany: High Jinks Below Nelson's Column', *The Manchester Guardian* (26 October 1948), 3.

> Mounted police had to part the crowds watching the floodlit fountains in Trafalgar
> Square last night [...] the square was like a fair, with fountains changing from flame
> to orange, coloured lights flickering on the standards over the ceremonial stands,
> and a searchlight from the Admiralty picking out Nelson high in the sky. People
> climbed railings, they climbed the lions, and tried to climb the trees; some waved
> flags brought out from a former jollification.[13]

This account gives a strong sense of the spectacle enjoyed by a crowd still
recovering from wartime drabness, and is an echo of the wonder described
at the VE Day floodlighting, discussed in Chapter 3. Royal occasions were
also important diversions and moments of national feeling, and always
brought a crowd into the Square. The period between the end of the war
and 1965 saw several such occasions: Princess Elizabeth's wedding in 1947,
George VI's funeral in 1952, the new Queen Elizabeth's coronation in
1953 and Churchill's state funeral in 1965. While this last event was not
strictly royal, it observed many similar conventions, including a service in
Westminster Cathedral and the attendance of numerous heads of state and
every European monarch at the time.[14]

In the strictly-rationed post-war environment, Princess Elizabeth's
1947 wedding was relatively low-key, with an example of austerity that the
cake was made from ingredients given to the royal couple as wedding gifts.
Many people, however, looked forward to the spectacle, and one observer
reported that 'Trafalgar Square was so crowded that not a pigeon could
find foothold'.[15] In *The Times*, the wedding was described as a reminder
of national resilience:

> This is a nation struggling, impoverished, perplexed. It is also a nation rich in new-won
> glory, and the brave panoply of a royal wedding, the glass coach and the glittering
> breastplates, the cheering multitudes and the pealing bells, proclaim that the glory
> is untarnished by the labours which are the aftermath of its winning.[16]

---

13 'Royal Sightseers: Drive last night round Trafalgar Square', *The Manchester Guardian*
   (25 October 1948), 5.
14 Geoffrey Best, *Churchill: A Study in Greatness* (London: Hambeldon and London,
   2001), 326.
15 Kynaston, *Austerity Britain*, 244.
16 'The Human Hope', *The Times* (20 November 1947), 5.

For the funeral of George VI, the crowd, 'marked by quiet, middle-aged women and policemen draped with water-proof capes', 'clustered more deeply in Trafalgar Square' than elsewhere along the processional route.[17] An image of the Square taken by Henri Cartier-Bresson shows dozens of solemn-faced people atop one of the lions, craning their necks for a better view of the cortege.[18] As with his daughter Elizabeth's wedding, *The Times* again drew 'lessons' about national character from the funeral: 'the nation, no less than the family, is a unit from which [the King's subjects] could not be cut off without the loss of their bloodstream'.[19]

The coronation of Elizabeth II over a year later saw people sleeping out in the Square to secure a good vantage point for the passing procession. Newspaper reports of the night before the Coronation described 'the morass in Trafalgar Square', where 'the crowds had settled down in solid rows that extended fifteen deep at their maximum'.[20] One report tapped the history of the Square to track national history, comparing the 1953 event to the coronation of her grandfather George V in 1911, again focusing on Trafalgar Square where people had camped through the night to secure a good seat: 'on the Nelson Monument they sat tightly in five tiers encrusting the crouching lions, and standing shoulder to shoulder against the plinth'.[21]

Winston Churchill's state funeral, for which he had requested 'plenty of bands', can be understood as the de facto end of British imperialism.[22] Prime Minister Harold Wilson began his tribute in *The Times* by reminding readers of Churchill's early imperial career 'in the Queen's service in the 4th Hussars, into the prisoner-of-war compound – and out of it; he charged

---

17 'Through Still Streets; The Procession's Way; Touch of Colour in Sombre Scene', *The Manchester Guardian* (12 February 1952).

18 In Hargreaves, *Trafalgar Square*, 64.

19 'Epilogue', *The Times* (16 February 1952), 7.

20 'Vast Crowds Encamp on Royal Route: Unprecedented Night Scenes in West End Streets', *The Manchester Guardian* (2 June 1953), 1.

21 'Coronation Vigils: Memories of 1902 and 1911', *The Manchester Guardian* (1 June 1953), 6.

22 Best, *Churchill*, 326 and Webster, *Englishness and Empire*.

with the Lancers and Omdurman, he was one of the first into Ladysmith'.[23]
The funeral was reported as crowded but muted, with the silence of the
'nine-deep crowds' in the Square broken only by the 'relentless thud of the
slow-marching feet' of the procession.[24] Official representatives came to
mourn from 111 countries, including 'four Kings and a Queen', demonstrating Churchill's international stature as a wartime leader.[25]

These events all spoke of British imperialism. For example, Cannadine
points out the imperial symbolism in Elizabeth's coronation ceremony –
the dominions' emblems on her dress, the presence of Commonwealth
and colonial troops – as well as her explicit reference to the uncertainty
of the transition to a post-colonial Commonwealth: 'I am sure that this,
my Coronation, is not a symbol of a power and a splendour that are gone,
but a declaration of our hopes in the future'.[26] A similar ambivalence had
been evident during the 1951 Festival of Britain. The *British Pathé* newsreel
of the King opening the Festival shows the royal procession moving 'across
Trafalgar Square and on into the Strand [...] through London's cheering
thousands'.[27] The slightly wistful accompanying commentary implies the
limits to British greatness: 'As a people building ourselves into a great
nation, we made our mistakes. Too often these alone are remembered now.
It is good therefore that we should remember that which is great in our
achieving'. This can be interpreted as a reference to the Empire, as during the
Festival itself there was limited mention of the Empire of Commonwealth,
'a politically sensitive and ambiguous topic' in 1951.[28]

Churchill's funeral in 1965, then, evoked a fading imperial past, embodied by the man himself. The *Weekend Telegraph* published the insert 'Last

---

23   'We Are Marking the End of an Era', *The Times* (25 January 1965), 8.

24   'The great from 110 nations gather to mourn', *The Observer* (31 January 1965), 2.

25   'Four Kings and a Queen among leaders from 113 nations', *The Times* (29 January
     1965), 15.

26   Cannadine, *The British Monarchy*, 150.

27   *British Pathé*, 'Special: The King Opens Festival 1951' <http://www.britishpathe.
     com/video/special-the-king-opens-festival> accessed 18 April 2013.

28   See Becky Conekin, *'The autobiography of a nation': The 1951 Festival of Britain*
     (Manchester: Manchester University Press, 2003), chapter 7.

Journey and Farewell to Greatness' with photographs of his funeral procession and text that described his military life and experiences overseas. Beyond his role as a hero of the Second World War, 'Churchill was credited for speaking for the nation on a deeper level – as an imperialist who had loved the British Empire, as a historian who had celebrated the great deeds of nation and empire [...]'.[29] The feeling reported at Churchill's death about Britain's 'diminished identity' implied not only nostalgia for the end of empire, but an uncertainty about new relationships with Commonwealth states.

Race and politics in Britain: The Anti-Apartheid Movement

One visible manifestation of the changing post-imperial relationship was Commonwealth migration, personified in 1948 by the arrival of 492 Jamaicans at Tilbury, less than thirty miles down the Thames from the Square, in the *Empire Windrush* on 22 June. Fryer describes this event as the arrival of much-needed workers: 'Officialdom, at both government and local levels, moved swiftly to make the Jamaicans feel welcome and find them accommodation and work'.[30] In the House of Commons, discussion of the new arrivals concerned their housing and employment, with some misgivings that the West Indians were not being treated as well as they should be, given that they were 'British subjects emigrating to other parts of the Commonwealth'.[31] These reports demonstrate a sense of responsibility felt by the British government towards these migrants, even though there were some concerns about migration on a scale too large for British society or industry to accommodate. Britain in the 1940s had a very small population of visible minorities, mostly in the port cities of Cardiff,

29　Webster, *Englishness and Empire*, 217.
30　Fryer, *Staying Power*, 372.
31　House of Commons Debate, 24 June 1948 vol 452 col 1555–6.

Bristol and Liverpool, and London's population was predominantly white and English-born. In the 1950s, 'half of Britain's white population had never even met a black person [...] more than two-thirds of Britain's white population, in fact, held a low opinion of black people or disapproved of them'.[32] Half of this prejudiced population were deeply so, and violence or more subtle forms of discrimination were common experiences for most black people in this period. Against this domestic background of increased Commonwealth migration, and ambivalence and even hostility towards migrants, the anti-apartheid movement took root in Britain.

Apartheid had been formalised in South Africa following the 1948 election of a Nationalist Party government that created legislation that would enable the systematic control and disadvantage of black South Africans. In Britain, opposition groups were bolstered by South African refugees and émigrés who helped to organise protests such as boycotts of South African goods.[33] In 1962, South Africa withdrew from the Commonwealth as a result of its unwillingness to change its apartheid policies. Despite Harold Macmillan's efforts to support South African membership, there was evidence of British government uncertainty about whether it weakened or strengthened the Commonwealth.[34] For many on the British left, South Africa's apartheid regime was a disgusting system for which Britain was at least partially responsible. In 1959, a campaign to boycott South African goods was organised, a tactic with 'wholly South African origins', where it had a long history and some success.[35] The next year, the boycott was on a wider scale and better organised, with a broader support base that included celebrities, academics and city councils, especially port cities with long-standing black communities.[36] Calling for a month-long consumer boycott of South African goods, timed in part to coincide with the spring arrival of fruit, a public rally was held in Trafalgar Square on 28 February 1960.

32   Fryer, *Staying Power*, 374.
33   Fieldhouse, *Anti-Apartheid*, 4–6.
34   Fieldhouse, *Anti-Apartheid*, 38–39.
35   Christabel Gurney, '"A Great Cause": The Origins of the Anti-Apartheid Movement, June 1959–March 1960', *Journal of Southern African Studies*, 26/10 (2000), 123–144.
36   See Gurney, 'A Great Cause'.

The Square was an obvious choice of location because of its longstanding history as a protest site, but also because South Africa House is located on the Square. During this period, CND was also using the Square for annual rallies, as discussed below, and the anti-apartheid campaigners sought their help in planning their initial rally. On the day of the protest, 'widespread publicity resulted in an estimated 2,000 people marching from Marble Arch to Trafalgar Square'.[37] According to *The Times* 'It was a bigger demonstration that many had expected. Some officials put the attendance as high as 15,000, but other estimates were nearer 6,000. About 2,000 took part in the march [from Marble Arch...]. Several in the procession looked as though they might have had previous experience of marches – there were some nuclear disarmament badges'.[38]

The February boycott rally was followed the next month by a much larger demonstration which saw the anti-apartheid movement really begin to take shape. On 27 March, protesters rallied against the 'Sharpeville Massacre', in which police fired on an unarmed crowd in a black township south of Johannesburg. Sixty-seven people were killed and 186 injured as they protested against the introduction of 'pass laws' which were designed to control the mobility of black South Africans.[39] The Trafalgar Square reaction to the shootings was described in *The Guardian* as 'the largest and most impressive demonstration to be seen in London since the [CND's] Aldermaston march last Easter'.[40] Organisers estimated that 15,000 people attended this rally, which benefited from arrangements for a march that had already been planned by the progressive group Christian Action. Although some arrests were made, *The Times* described an atmosphere of 'almost exemplary orderliness [...] it was only when the meeting was over and the square emptying that any disturbance occurred'.[41] *The Guardian* coverage

---

37 Roger Fieldhouse, *Anti-apartheid: a history of the movement in Britain: a study in pressure group politics* (London: Merlin, 2005), 18.
38 'Nine arrested After South Africa Boycott Rally', *The Times* (29 February 1960), 10.
39 Fieldhouse, *Anti-apartheid*, 20.
40 'Thousands speak out against Sharpeville: Massive protest in Trafalgar Square', *The Guardian* (28 March 1960), 1.
41 'London Rally Votes for Continuing Boycott', *The Times* (28 March 1960), 3.

suggested that the meaning of the Square might have been rewritten by the protest: 'the battles of Trafalgar Square might be remembered more vividly 50 years hence than the Battle of Trafalgar itself'.[42] Foreshadowing the debates over national identity prompted by the next century's Fourth Plinth scheme, the reporter went on to describe the earnestness of the young protesters, asking 'might some sculptor of the future feel moved to make a new sort of monument, a statue of a non-fighter in a duffle-coat?'[43] Catalysed by this support, the Anti-Apartheid Movement was officially established on 1 April 1960.

The comparison to an earlier CND rally, and mention of seasoned 'campaigners' wearing CND badges, points to the larger picture of connections among different left-oriented movements at the time, not least due to their use of the same venues, which built up layers of spatial meaning that other groups could then activate. Gurney argues that in the early years of the AAM, apartheid 'became one of the causes taken up by a network of organisations and individuals involved in a ferment of activity on three inter-linked issues: the anti-colonial struggle; peace and nuclear disarmament; and opposition to endemic and growing racism in Britain'.[44] The end of empire and fading British global power was mixed with public concerns about migration from the former colonies and insecurity about Britain's ability to protect herself from the larger Cold War. These three aspects of Britishness in the 1960s point to a transformation in the way the nation was understood. In terms of anti-apartheid activities, although Britain was no longer an imperial power, many on the left still felt a burden of culpability for populations it had once ruled, and systems it had bequeathed to independent states. One analysis of the 1960 demonstration against the Sharpeville Massacre, for example, argued that 'the British people are driven to protest [...] because they feel, however remotely, responsible [...]'.[45] In choosing to express their opposition to apartheid in Trafalgar Square, protesters broadcast their message from the very centre of the Empire they could not forget, and which they still felt was somehow theirs.

42   'London Letter: They Also March', *The Guardian* (25 March 1960), 12.
43   *Ibid.*
44   Gurney, 'A Great Cause', 128.
45   'Not Smug', *The Guardian* (27 March 1960), 16.

The discomfort on the left about responsibility for colonial systems of rule found a parallel on the right in concern about the impact of decolonisation on British society. In Trafalgar Square, this was manifested in 1960 in organised heckling from young men from Mosley's Union Movement who, according to varied press reports, called out 'keep South Africa white' and 'keep Britain white'.[46] Mosley himself spoke at a rally in the Square in July 1962 in support of the National Party of Europe. A film made by *British Pathé* shows a rowdy crowd barely held back by police, who eventually break through the cordon to attack the plinth from which Mosley's supporters are speaking. According to the newsreel's commentary, the rally had been allowed to occur on the basis of 'free speech', with authorities suggesting that 'those who can't bear to listen to fascist propaganda would do better to stay away'. However, continued the narrator, 'Many in the crowd had long memories, recalling Mosley's anti-Jewish attitude before the war and his admiration for Hitler. For fascism, the British people have nothing but loathing and contempt'. Mosley blamed communists and promised to hold further rallies.[47] More generally, the organised and politicised racism during this period was typified by Enoch Powell's address to a meeting of Birmingham's Conservative Association in 1968. In his well-known 'Rivers of Blood' speech, Powell predicted a breakdown of British identity due to an enormous increase in 'Commonwealth immigrants and their descendants', describing this process as a 'total transformation to which there is no parallel in a thousand years of English history'.[48] For Mosley and Powell, racial issues were national issues. The Square was one site in which this national contest took place.

The AAM continued to enjoy a broad support base throughout the 1960s. A Trafalgar Square meeting on 3 November 1963 was met with messages of support and participation from universities, trade unions, Members of Parliament and prominent people including racing driver Stirling Moss, who sent the message: 'Let me say how sympathetic I am

---

46 'London Letter: Trouble in the Square', *The Guardian* (27 June 1960) and 'Trafalgar Square Incidents', *The Guardian* (1 July 1960).

47 *British Pathé*, 'Crowd Wrecks Mosley Rally, 1962' < http://www.britishpathe.com/ video/crowd-wrecks-mosley-rally> accessed 16 April 2013.

48 'Rivers of blood', *The Telegraph*.

towards the movement'.[49] As with demonstrations by previous users, the role of the Square in garnering publicity for the cause was recognised even by those outside the country. President Nkrumah of Ghana expressed his support of a rally in 1963:

> It is my earnest hope that the disgust and indignation of the British people expressed in various protests and demonstrations and parriculaly [sic] in to-day's march to Trafalgar Square, will help to focus attention on this evil policy of Apartheid [...]. I am with you in this march and Africa marches with you.[50]

In another letter of support for the same rally, President Julius K. Nyerre of Tanganyika identified the important role that the British people had to play in opposing apartheid: 'Britain is a major supporting power as far as the South African economy is concerned and could thus be an important factor in making it difficult for the South African Government to continue its racial tyranny'.[51] As the letters from these African leaders demonstrate, the AAM's use of the Square included an international audience, and its supporters defined Britishness to include British responsibilities to Commonwealth states. It is possible that imperial symbolism in the Square subtly reinforced this connection. More certainly, proximity to South Africa House made the Square the perfect place to mount visible protests. The Square's established history as a site of primarily left-wing protest and its public prominence, also enhanced the AAM's demonstrations.

Almost a decade later the Square took on another role as it was transformed into a symbolically South African site. To commemorate the ten-year anniversary of Sharpeville, on 21 March 1970 a graphic re-enactment of the massacre took place in the Square, described in the *Anti-Apartheid*

49   Letter from Stirling Moss to the AAM, November 1963, Shelfmark MSS AAM 2029, Anti-Apartheid Movement Archive, Bodleian Library of Commonwealth and African Studies, University of Oxford, 4.

50   Letter from President Nkrumah of Ghana to the AAM, Shelfmark MSS AAM 2029, Anti-Apartheid Movement Archive, Bodleian Library of Commonwealth and African Studies, University of Oxford, 6.

51   Letter from President Julius K. Nyerre of Tanganyika to the AAM, Shelfmark MSS AAM 2029, Anti-Apartheid Movement Archive, Bodleian Library of Commonwealth and African Studies, University of Oxford, 7.

News: 'The "police" took aim, the shots rang out and people in the crowd, many of them South Africans in exile, fell to the ground just as the real victims did at Sharpeville ten years before'.[52] Roger Jackson's striking image of this event shows a participant, his face drawn and haunted, carrying an effigy of a child's body. This emotional event drew a direct line to Sharpeville in 1960, linking the Square through its use to another time and place, and enriching the site's meaning by constructing it as South African. It also reiterated the sense of responsibility that British protesters felt for people in former colonies by using the imperial symbolism of the Square and South Africa House to reinforce the historical connection. Furthermore, as was often the case with the Square, the potential for publicity was an attractive aspect of the location. The 1969/1970 Annual Report of the AAM pointed out that as a result of the reenactment of the Sharpeville Massacre in Trafalgar Square, 'Large pictures of the re-enactment made the front pages of three Sunday newspapers and it received extensive publicity of radio and television'.[53] However, for some, this form of protest had lost its power. In a letter to the editor of *The Times* in 1970, Michael Langley wrote that 'the re-enactment of the Sharpeville massacre in Trafalgar Square under the windows of South Africa House, has long been a dreary inevitability'.[54] Tony Susman, a member of one of the groups that had organised the re-enactment, took issue with Langley's 'rather silly and irrelevant letter', describing Sharpeville as 'a manifestation of an inhuman, brutalizing, and degrading social and economic system [...] supported [...] by the British Government and British finance and industry [...]. We re-enact Sharpeville because in South Africa nothing has changed [since the massacre in 1960], apartheid is as rampant as ever, and Britain bears a very heavy responsibility'.[55] Here the sense of culpability felt by members of the AAM for Britain's imperial past is explicit.

---

52   *Anti-Apartheid News* (21 March 1970), 1.
53   AAM Annual Report, 1969/1970, Anti-Apartheid Movement Archive, Bodleian Library of Commonwealth and African Studies, University of Oxford, 23.
54   Michael Langley, 'Sharpeville Massacre' in 'Letters to the Editor', *The Times* (25 March 1970).
55   Tony Susman, 'Sharpeville Massacre' in 'Letters to the Editor', *The Times* (2 April 1970).

Perhaps more than other protests, the AAM's use of the Square demonstrates Massey's notion of a 'global sense of place' in which places are open and hybrid, always changing and inextricably linked to other places and times.[56] In this case, the Square linked London with the political and social inequality in South Africa through its imperial symbolism and that of South Africa house, as well as through calls for the empowerment of social groups that had echoes in the site's previous uses. Webster argues that by the mid 1960s, 'the Commonwealth, previously associated with a world-wide community – the "people's empire" of a racial community of Britons, or the "people's empire" of a multiracial family of nations – acquired a wholly new meaning: domestic problem.'[57] This change was driven by white British reactions to 'coloured immigration', a problem that was often constructed in terms of the migrants' difference from the white working-class communities they joined.

At the same time, as the evidence above suggests, the progressive and middle-class left was exercised about the brutality and injustice of the apartheid system, and demonstrated against it. These two aspects of the British relationship to former colonies point to the complexity and dynamism of the social change that the country was undergoing only two decades after the end of the war, and also hint at the range of ways that 'Britishness' could be used to understand this change. The domestic reaction to large-scale international changes included the ongoing use of Trafalgar Square for rallies and demonstrations, just as in previous decades. The Square's role as a British site that included inescapably imperial symbolism, while never explicitly mentioned by the AAM, was implicit in its use, which sought both to demand change to the apartheid system in South Africa and to demonstrate a sense of responsibility for that system.

56   See Massey, 'A global sense of place'.
57   Webster, *Englishness and Empire*, 149.

Campaign for Nuclear Disarmament

Apartheid was not the only international issue that drove demonstrators
to the Square. In 1956, Anuerin Bevan addressed a large crowd during the
Suez crisis, describing the Government policy as one of 'bankruptcy and
despair' and arguing that Britain was part of a larger international system:
'we are stronger than Egypt, but there are other countries stronger than
us'.[58] The speech hinted at Labour's internationalism, which saw itself as
connected to the international community, and sought to orient British
foreign policy towards cooperation with other states, including through
international organisations such as the United Nations.[59] By the late 1950s,
the development of nuclear weapons and the beginning of the Cold War
saw the Campaign for Nuclear Disarmament use the site in several well-
attended rallies. CND was established in 1958, and one of its first organised
protests was a march from Trafalgar Square to a US base in Aldermaston,
near Reading, about eighty km west of the capital. Meticulous planning
for this event was evident in the 'Instructions to Stewards and Marshals',
down to the details of cleaning up after the marchers: 'The success of the
march can be seriously jeopardised by failure to deal with the problem of –
LITTER! Each Marshal is asked to call for one volunteer from his Group to
be responsible for ensuring that litter is *not* left at Halting Points. A Litter
Van will, if possible, be provided, with sacks'.[60] These details show a careful
level of organisation (and a concern with tidiness) that typified the CND
in many of its activities. By the next year, the route had been changed to
go from Aldermaston to Trafalgar Square. One reason for this was made
explicit in a pamphlet entitled 'why we are marching', which explained:

58   A recording of this speech is available on *The Guardian* website 'Great Speeches
      on the Twentieth Century': <http://www.guardian.co.uk/greatspeeches/
      bevan/0,,2060115,00.html > accessed 6 June 2013.
59   See Paul Corthorn and Jonathan Davis, *The British Labour Party and the Wider
      World: Domestic Politics, Internationalism and Foreign Policy* (London: Tauris, 2008).
60   London School of Economics, Campaign for Nuclear Disarmament Collection
      (LSE/CND), 4/1 'Instructions to Stewards and Marshals', 1958, 6.

'LONDON is our capital city, where Government and Parliament make decisions that spell life and death for us all and where we must raise our voices if we want a change.'[61] Trafalgar Square's proximity to the location of government and history of use almost certainly enhanced their demands. The 1959 rally was very well-attended, with one press report asking if it was the 'biggest protest of the century?' and estimating attendance at 20,000.[62] As the 'crowds packed tight in Trafalgar Square', leaders claimed that 'in all its long history Trafalgar Square has never seen a demonstration of this size [... and] intensity.'[63] In the House of Commons, MP Frank Beswick agreed on the march's significance:

> the demonstration which culminated on Easter Monday in Trafalgar Square was the largest single political demonstration which has taken place in this country since the war and probably of all time, and that it was not only its size but its representative character which was so impressive [...]. Is not this demonstration of public opinion, at a time when all the major political parties are complaining of political apathy, something which we should be very unwise to ignore?

Apparently, however, Home Secretary Rab Butler was not convinced, replying that the CND did not 'represent either the opinion of the majority or the best interests of this country.'[64] In 1960, the size of the Easter rally was yet bigger, with even the police confirming that it was the biggest demonstration ever held in the Square. *The Times* was not as committed, describing it as 'the biggest demonstration seen for a long time'. However, it did link the protesters to a breadth of national support for CND's aims:

> What was perhaps most striking was the great variety of their home towns, combined with the unity of their appearance. A roll-call of English place-names seemed to be displayed on the banners [...] similarly there were groups of organisations in the column which together suggested a great breadth of opinion.[65]

---

61   LSE/CND/4/1, 'Why We Are Marching', 1961.
62   'Bomb Marchers Fill Trafalgar Square: Biggest Protest of Century?', *The Manchester Guardian* (31 March 1959), 12.
63   '10,000 Marchers in Trafalgar Square', *The Times* (31 March 1959), 4.
64   House of Commons Debate, 14 April 1959 vol 603 cc 809–811.
65   *The Times*, '60,000 in Bomb Protest Demonstration', 19 April 1960, 10.

For the speakers addressing the 40,000 attendees, the implication was that 'nobody can ignore us now'; it identified the importance of the event for showing the movement to itself:

> Whatever comes of the speeches in the square, or the political impact of the march, at least it did one big thing for the marchers themselves. Any one of them – the suburban housewife, the pastry cook, the railwayman, the Nigerian student, the actress – now know that an extraordinary variety of people feel the same way as they do. They have seen them.[66]

Here, being in the Square on Easter Sunday 1960 made participants not only evident to each other, but to government and, though the press, the wider British public. As I have been arguing throughout this book, this is one of the key ways that Trafalgar Square has made its mark on national identity: by providing a space for different groups to stake a claim in the nation, populating their 'imagined communities' with tangible numbers of supporters and making them visible within the nation. The Square's central location and history of protest have aided this process.

The rallies continued to grow in size and popular awareness, although in 1962, the movement that was too big to be ignored was caught out by its opponents. CND did not use the Square that year, the reason why revealed in a *Guardian* article, 'Right-wing steals a march: CND misses Trafalgar Square':

> 'The "Keep Britain Great" Committee has bagged the site instead' the Committee's leader, Mr W.H. Gleaves said that he and his friends were 'sick and tired of the weirdies getting all the publicity'. So he booked Trafalgar Square for Easter Monday 'It has been rumoured in London that someone in CND's office simply forgot to renew last year's booking [... and] nobody is allowed to book Trafalgar Square while it still has a demonstration pending in the same place.[67]

66 'Journey's End for 40,000 Marchers: Anti-nuclear demonstrators in Trafalgar Square', *The Guardian* (19 April 1960), 1.
67 'Right-Wing steals a march: CND misses Trafalgar Square', *The Guardian* (9 January 1962), 1.

This potentially embarrassing oversight was cast in positive terms in an article from the CND's newsletter *SANITY* entitled '20,000 – and still growing'. In it, the CND National Secretary Peggy Duff commented: 'It's lucky we couldn't get Trafalgar Square for tomorrow [Easter Monday] [...]. We would never get them all in'. In the same publication, however, Trafalgar Square was praised in an article linking the CND's campaign of public demonstrations to a long history of British political protest, including one captured in the accompanying photograph captioned 'Trafalgar Square, focus of protest: a socialist rally 80 years ago':

> CND fits well into the picture of the pressure groups which have been the plague of Governments, and at the same time largely responsible for the better things in life in modern Britain [...]. So you who are marching today are flowing in a great tradition of those who worked to make our land a juster and fairer isle. But there is one great difference. Those in the past had time on their side. They could wait ten years or more until their hopes were realized. But we haven't time to let them take over our ideas and pretend they thought of them first. For us, it is a question of NOW OR NEVER.[68]

Like others before them, the CND used the previous history of protest in the Square to help legitimise and augment the significance of their own use of it. By 1965, the central office appeared to have learned the lesson of booking the Square early, although the starting point of the march had shifted from Aldermaston to High Wycombe. A flyer for that Easter's rally confidently proclaimed:

> So we are MARCHING AT EASTER from the Headquarters of Britain's Bomber Command [...] HIGH WYCOMBE past the Headquarters of the American Third Air Force [...] RUISLIP to the Headquarters of the British Government, WHITEHALL AND TRAFALGAR SQUARE [...] TAKE THE ROAD TO DISARMAMENT – WITH US AT EASTER![69]

Here, the road to disarmament, the CND's political goal, was glossed as the marching route that ended in Trafalgar Square. Thus, like other groups before them, in both symbolic and material terms, the CND conceptualised

---

68   'Tony McCarthy on Britain's Story of Protest', *SANITY*, Easter Monday 1962, 4–5.
69   LSE/CND/4/5, 'Administrative papers and correspondence', 1965.

the Square as a national space in which they could stake their claims for national political change. By putting the Square on the 'road to disarmament', CND made the Square central to their version of the nation.

Overall, the picture that emerges by 1965 is one of a well-coordinated organisation with ample experience in mounting multi-day protests. There was a very detailed organisational structure and a hierarchical system of marshals, each with their own differently coloured armbands. There were calls for volunteers for medical teams ('chiropodists and nurses are as useful as doctors'), as well as child care, litter teams, and transport, and provision for buckets, canvas screens and portable toilets. The routes and timings were worked out in careful detail. The minutiae of the planning suggest that an orderly image was important to CND, helping to make respectable their campaign, with its many middle-class supporters and largely peaceful means of political pressure. However, by the late 1960s CND's formal membership was in decline, with left-wing causes such as protest against the Vietnam War attracting its support base. One response to this by the CND was to try and show its relevance to the proxy wars between the US and the USSR, and in 1967 it invited 'Madame Thi Bihn of the National Liberation Front of South Vietnam' to appear at a rally in Trafalgar Square:

> The high point of the weekend demonstration will be the Easter Monday Rally in Trafalgar Square, London at which there will be an audience of at least 25,000. This will be the occasion for the major political statements, where the Vietnamese representatives will explain the nature of their struggle. This Rally will have wide publicity in the press, radio and television. The previous meetings and activities enhance the importance of the final rally by contributing to the publicity build up, but the full import of the message to the British people should be reserved for the final rally.[70]

This internal memo clearly identifies the public impact that it was hoped the use of the Square would have. Despite its drop in prominence, the CND still looked to Trafalgar Square as an important site to make a national statement. By the end of the 1960s, however, the CND was in decline as an organisation, despite the high profile it had enjoyed at the beginning

70   LSE/CND/4/24, 'Invitation to Madame Thi Bihn', 1967, 1–2.

of the decade. The early 1970s saw a return to large labour protests in the Square that highlighted the domestic challenges facing the country, as well as the international ones.

Anarchy in the UK? 1970–1990

The 1970s have been characterised as a period of national conflict, even anger, a period of 'extraordinary cultural and social flux'.[71] Conflict over the status and identity of Northern Ireland, Britain's changing relationship with Europe, the Commonwealth and the United States, and industrial relations were a feature of national life. Disagreements between unionised industry and the Government had a particularly acute effect on Britain. An important touchpoint in the early 1970s was the proposed Industrial Relations Bill. February 1971 saw a trade union rally in Trafalgar Square that drew a massive crowd – the Trade Unions Council estimated attendance at 140,000 – protesting against the Bill. The demonstrators came from around Britain, mustered in Hyde Park and marched to Trafalgar Square in a show of solidarity. The rally was so large, however, that TUC General Secretary Vic Feather led the crowd from the Square to the Embankment so that everyone could be accommodated. The march was peaceful, but the numbers caught everyone by surprise: one police officer observed 'God only knows what we would have done if Feather hadn't started to lead them off to the Embankment'.[72] Feather himself described the rally as 'the

---

71   See Richard Weight, *Patriots: National Identity in Britain 1940–2000* (London: Macmillan, 2002); Mark Garnett, *From Anger to Apathy: The Story of Politics, Society and Popular Culture in Britain since 1975* (London: Vintage, 2007); and Dominic Sandbrook, *State of Emergency. The Way We Were: Britain, 1970–1974* (London: Allen Lane, 2010), 12.

72   '125,000 march against Bill', *The Guardian* (22 February 1971), 1.

biggest organized demonstration ever to take place in Britain'.[73] Like many protesters in the Square before and after them, the organisers appeared to hope that the huge turnout would influence Government policy and make their demands irresistible. The rally was mentioned repeatedly in the House of Commons debate on the Industrial Relations Bill the following week. Those who supported the bill claimed the turnout was a result of widespread confusion about the proposed law's implications, and those who did not claimed it represented popular outrage. Given the attention that the Trafalgar Square protest received in Parliament, the demonstration can be said to have bolstered the campaign to 'kill the bill', maintaining the momentum that contributed to the eventual legislative repeal under Harold Wilson's Labour government in 1974. The show of strength by the unions in the Square may also have had the unintended consequence of portraying them as 'too powerful', and thus helping to justify Thatcher's anti-trade union legislation in the 1980s.[74]

The winter of 1973 saw miners' strikes that prompted rolling power cuts, the declaration of a state of national emergency and the introduction of a three-day working week to save fuel. Despite what Vic Feather later described as 'the biggest demonstration in recorded history', unions were not able to avoid further conflict with the government.[75] The Square's public prominence, however, did allow the TUC to display its strength and breadth of support to both the Government and the broader public, even if this did not help advance its political goals. Certainly the rally in the Square would have contributed to the effect this conflict had on Britishness in the 1970s; as Weight explains it:

73   Peter Weymark, '100,000 march in peaceful protest against unions Bill', *The Times* (22 February 1971), 1.

74   House of Commons Debate, 23 February 1971 vol 812 cc323–91. Many thanks also to Paul Ward for pointing out some of the political ramifications of the 1971 TUC rally in the Square.

75   'Mr Feather says anti-Bill protest "biggest in history"', *The Guardian* (22 February 1971), 18.

Primarily, it gave the progressive remaking of Britishness a sharp jolt by showing how much class conflict remained a feature of national life. For all the advances made since the 1950s towards the creation of a more meritocratic society, institutionalised inequality of opportunity was still rife [... despite complexities] in the popular imagination the conflict was seen to be between the working and middle classes.[76]

Meanwhile, on the Cold War front, by the mid-1970s it appeared that public pressure against nuclear weapons was no longer necessary because 'it was widely assumed that the superpowers had entered a new era of sanity' that rested on a gradual process of disarmament.[77] This changed again in the 1980s, however, when newly elected US President Ronald Reagan adopted a much more aggressive position towards the 'evil empire' of the USSR than his predecessor Jimmy Carter. Despite a large drop in CND membership through the 1970s, Reagan's 'unnerving policy of nuclear brinksmanship' saw a drastic increase in CND by 1983 to around 90,000 fully paid members.[78]

However, even given more than a decade of agitation, including some very large rallies in the Square, it is not clear that CND had any influence over British nuclear policy. CND's activities can be understood as part of a long history of popular protests to government policy, with the Square providing a venue for alternative, vernacular versions of British feeling and identity to be expressed. A national space, the Square helped an alternative Britain to identify itself and remind the government that minority concerns still had a place. The size of some of these rallies, and the granting of official approval to use the site, suggests that the expression of alternative views was possibly viewed as a safety valve by government. Certainly the notion of the right to free speech that had underpinned protests in the Square in the past was still relevant in this period. While the AAM and CND may not have been entirely mainstream organisations (although they did enjoy large memberships at some times), they were part of the mix of the British political and social landscape, and they used Trafalgar Square to make their national claims.

76   Weight, *Patriots*, 524–525.
77   Garnett, *From Anger to Apathy*, 121.
78   *Ibid.*, 123.

By the 1980s, the AAM was regularly using the Square for protest rallies. As in earlier demonstrations, the connection between London and South Africa was an ongoing theme. For example, at a rally on 14 March 1982, MP Tony Benn said 'the roots of apartheid are here in the City of London where great money and profit is made out of the exploitation of black workers in South Africa'.[79] One of the biggest rallies in this period was in 1984 in response to Prime Minister Margaret Thatcher's invitation of South African Prime Minister P.W. Botha to visit London. The AAM's 1983/1984 Annual Report describes how important this event was to the movement:

> The AAM was faced with the immediate task of how to mobilise the widest possible opposition to the visit and in particular to succeed with its pledge to organize the largest-ever Anti-Apartheid Movement demonstration on the streets of London.[80]

According to the Annual Report, the rally drew together a range of participants, including local black and anti-racist organisations, the Greater London Council (including its leader Ken Livingstone), the Labour party, the Communist party, Plaid Cymru, the National Union of Students and the Trade Union Council:

> Saturday 2 June saw an estimated 50,000 people gather on the streets of London to protest against the visit. The march assembled in Hyde Park, where there was an hour-long rally before moving off along Piccadilly to Trafalgar Square, down Whitehall, where a letter was handed in at Downing Street, and on to Jubilee Gardens on the south bank for another rally and a festival.[81]

Speakers cast the relevance of apartheid to the British population in terms of British race relations: MP Roy Hattersley, for example, said that a positive reception for Botha and 'his beliefs in the innate superiority of the white race is openly to affront the black and Asian British', reinforcing the link between the movement against South African apartheid and race

79   *Anti-Apartheid News* (March 1982), 12.
80   AAM Annual Report 1983/1984, 11.
81   *Ibid*, 12.

relations in Britain.[82] In the same newspaper report, the AAM claimed that the 25,000-strong rally was the 'biggest demonstration in its 25 year history' and went on to report its impact in its July–August Newsletter:

> The deep anger and resentment aroused by his [Botha's] visit throughout Britain, and in all sections of the community, was reflected by the tens of thousands of demonstrators who gathered in London on 2 June. They caused Mrs Thatcher when she met PW Botha not only to condemn apartheid but to echo a number of specific concerns regarding issues on which the AAM has consistently campaigned.[83]

Here, the AAM argued that the Trafalgar Square rally had an influence on the meeting and potentially helped to shape British policy towards South Africa. However, according to *The Times*, 'For most politically-aware blacks the tour has been an unmitigated disaster, conferring on Mr Botha an international respectability which in their eyes he has done nothing to deserve'.[84] In the House of Commons, the MP for Liverpool Garston, Eddie Loyden, described the visit by Botha as 'an insult to the majority of people in Britain and an insult to most Commonwealth countries', while the Foreign Secretary, Geoffrey Howe, defended it as an opportunity of 'making plain to the South African Government our strongly held views on many South African questions', including describing the government's views on apartheid as also 'strongly held' without specifying what they were.[85] All these accounts show that the question of South African apartheid was a matter of public concern in Britain, and that the use of the Square was part of a larger debate about Britain's role in creating it and responsibility for helping to end it.

Even among those who opposed apartheid, there was disagreement about the best way to achieve this. The AAM, for example, had a strong notion of 'lawful' or legitimate protests, and had a broad national support base and tactics that included lobbying, economic boycotts and attempts to apply political pressure. This approach was at odds with the direct action

---

82   '25,000 join demonstration against Botha's visit', *The Guardian* (4 June 1984), 2.
83   AAM, July–August Newsletter, 1984, 4.
84   Michael Hornsby, 'Botha tour hailed as putting an end to pariah status' *The Times* (15 June 1984), 5.
85   House of Commons Debate, 27 June 1984 Vol 62 cc978–981.

favoured by the City of London AA Group (CLAAG), which drew many of its members from the Revolutionary Communist Group (RCG). The main source of the conflict between these two groups was over ideology, with CLAAG insisting that apartheid was inseparable from the inequities of capitalism, imperialism and racism, and that fighting it demanded a more active approach. In 1982, CLAAG began a non-stop picket of South Africa House in Trafalgar Square, which was 'followed by regular pickets over the next few years and then another non-stop vigil from April 1986 until Mandela's release in February 1990'.[86]

CLAAG's confrontational approach, inevitable conflict with police and disdain of the AAM's strategic links with the less radical Labour Party saw the AAM distance itself from the more radical organisation, withdrawing formal recognition in 1985. In one annual report it grumbled that the aggressive nature of CLAAG's pickets had 'resulted in a number of long-standing supporters of the Movement refusing to attend the pickets [...] shouting has been carried on in so aggressive and intimidating a fashion that passers-by have been deterred from accepting leaflets, and it has been felt that this has obscured the message of the demonstration'.[87] On the other hand, one new member of CLAAG was inspired to join upon hearing about the regular picketing of South Africa House; in the same report he said that he had previously 'thought of the AAM as an organization which talked and debated rather than one which engaged in concrete political action'. The more active tactics clearly had a motivating appeal for some people.

Police response to the pickets could be at least as aggressive as the protesters' tactics. In July 1984, the Metropolitan Police moved to ban the ongoing demonstration facing South Africa House on the basis of one police commander's interpretation of the Vienna Convention, which ruled that signatories must protect the dignity of foreign diplomatic missions.[88] This effectively denied demonstrators the de facto 'right to protest' which

---

86 Fieldhouse, *Anti-Apartheid*, 223.
87 AAM Annual Report 1985.
88 'Police officer tells court demo was banned to preserve embassy dignity', *The Guardian* (24 July 1984), 2.

had been particularly important in the Square itself. In a 10 June 1984 press statement, CLAAG described the police response to a group of eighteen demonstrators after the ban was introduced: 'Within five or six minutes a massive police cordon surrounded the peaceful picketers. They were all arrested without being told why and bundled into waiting police vans.'[89] A 6 July letter from campaigner Steven Kitson on 'recent events outside South Africa House' reported: 'Police from the coach pounced on a protester. He was assaulted and carried by four or five officers to a police van. He was later released without charge!'[90] The conflict between the AAM and CLAAG shows that disagreement over how best to use the site for protest was not just between the right and the left, or the government and inter-est groups, but occurred *within* broad movements. The Square was clearly valuable territory in terms of publicity and in demonstrating to decision makers that public sympathies supported the cause.

When Nelson Mandela was released from prison in South Africa in 1990, London was an epicentre of celebration for the end of his imprison-ment and the apartheid era. In May 1990, the *Anti-Apartheid News* reported on a major demonstration in Trafalgar Square:

> 25 March saw over 20,000 supporters in London for the Anti-Apartheid Movement's national demonstration [...] with a message from Walter Sisulu and Nelson Mandela [...] 'In our years on Robben Island and Pollsmor prisons, every attempt was made to break our spirit. Our imprisonment was aimed at isolating us from our people and the rest of the world. But we were never alone. You continued to inspire us from outside our prison walls [...]'[91]

This was vindication not only of the effort of anti-apartheid campaigners for the past thirty years, but also of the public prominence that the use of Trafalgar Square helped to build for protesters. Here, the Square was

89   Anti-Apartheid Movement, CLAAG Press statement, 10 June 1984, Shelfmark MSS AAM 502, Anti-Apartheid Movement Archive, Bodleian Library of Commonwealth and African Studies, University of Oxford.

90   Anti-Apartheid Movement, Letter from Steve Kitson, 6 July 1984, Shelfmark MSS AAM 502, Anti-Apartheid Movement Archive, Bodleian Library of Commonwealth and African Studies, University of Oxford.

91   *Anti-Apartheid News* (1990).

connected to other sites in a purposeful and political way, as it had been, for example, in the 1970 commemoration of the Sharpeville Massacre. It is hard to assess the contribution that the use of Trafalgar Square made to the British anti-apartheid campaign. Perhaps it is more useful to frame the use of the Square as an indication of how Britain (and London) was connected to South Africa through its use as a protest site, as in the Sharpeville Massacre re-enactments, or direct reference to the Square by various African leaders. The sense of responsibility felt by Britons that was expressed in the early years of the boycott suggest that this use of the Square showed the connection to former empire that continued to influence British national identity through the twentieth century and, as I will discuss in the next chapter, into the twenty-first.

Poll Tax Riots: 1990

On 31 March 1990, six days after the triumphant rally supporting Sisulu and Mandela, Trafalgar Square saw the most violent protest in Trafalgar Square since 'Bloody Sunday' in November 1887. The Poll Tax Riot began as a demonstration against the introduction of a new tax that was perceived as discriminatory towards the poor. A peaceful demonstration that began in the afternoon escalated into a violent confrontation between protesters and police, involving around 3,000 people.[92]

The riot developed from scuffles outside Downing Street, when sit-down protesters refused to disperse and police began to make arrests.[93] In the Square, scaffolding and workers' huts from a building site adjacent to the Square were ignited, neighbouring South Africa House was attacked, and 'bricks, bottles, scaffolding poles, oil drums and fire-extinguishers were thrown at police'.[94] A photographer who tried to escape from the

---

92   Hood, *Trafalgar Square*, 124.
93   '1990: Violence flares at poll tax demonstration', *BBC News* (31 March 1990) <http://news.bbc.co.uk/onthisday/hi/dates/stories/march/31/newsid_2530000/2530763.stm > accessed 18 April 2013.
94   David Butler, Andrew Adonis, and Tony Travers, *Failure in British Government: The Politics of the Poll Tax*. (Oxford: Oxford University Press, 1994).

Square said 'it was a crazy situation; in just three quarters of an hour it had turned from a fun family atmosphere into an angry, tense occasion. There was a mob on the streets'.[95] Despite official efforts to disperse the crowd, by early evening, parts of nearby Soho, Covent Garden and Charing Cross were strewn with broken windows and burned-out cars. 339 people were arrested, 374 police officers and 86 members of the public were injured, twenty of the forty police horses on duty were also injured and there were 250 reports of damage to property.[96]

The Poll Tax protesters effectively controlled the Square and its surrounding neighborhood for hours, gaining an important symbolic victory for the unofficial voices that had often used the site to stake their national claims. The protest was also part of a much larger non-payment campaign that saw the end of the Poll Tax; residents of Glasgow had been protesting since its introduction in April 1989, and at least 38 per cent of the registered Scottish population was subject to legal proceedings as a result of non-payment. The London riot, however, had UK-wide impact:

> The poll tax riots of March 1990 had helped to induce a change of government policy, and the participants could even claim they had contributed to Mrs Thatcher's departure from Downing Street. One reason for the success of this protest was its geographical location; the image of a burning car near Trafalgar Square proved [...] arresting for the national media [...][97]

In an echo of late Victorian concerns about mob rule and the degeneracy of the poor that influenced mainstream reporting of disturbances in 1886–1887, press coverage of the 1990 riot included reference to the protesters as 'an alienated British "underclass"', portrayed the police as victims of criminal hooligans and thugs, and emphasised an underlying criminality and conspiracy.[98] Again echoing historical rhetoric, in the House of Commons Sir John Wheeler MP described the police as '[defending] the principle

95   Catherine Pepinster, 'Shoppers hurt by rioting mob', *The Observer* (1 April 1990).
96   Butler, Adonis and Travers, *Failure*, 153.
97   Garnett, *From Anger to Apathy*, 329.
98   David Deacon, and Peter Golding, *Taxation and Representation: The Media, Political Communication and the Poll Tax* (London: John Libbey, 1994), 132–133.

of parliamentary democracy and freedom'.[99] Labour Party Deputy Leader Roy Hattersley also characterised the conflict as one between a democratic society and criminal forces:

> All democrats will combine in demanding the rooting out of the threat to our free society which was perpetrated by individuals and organisations who were responsible for the disgraceful scenes and conduct in the capital last Saturday.[100]

Rioters also saw their activities in terms of democracy and freedom. Accounts by rioters characterised the event as a 'class war', claiming that 'Trafalgar Square showed what was possible [...] that people were not prepared to take shit lying down and were able to organise resistance without leaders or parties'.[101] This interpretation was vindicated by the fact that, despite the violence, the Commission of the Metropolitan Police 'advised the Government not to over-react' to the riot on the basis that diverting protests to other locations would simply divert violence, and that 'it would be naïve to try to stop people demonstrating near the door of the Prime Minister' in nearby Downing Street where the first confrontation with police had happened.[102] This point of agreement between the rioters and police shows that the Poll Tax Riot was not merely a struggle over Trafalgar Square. It indicates acceptance that vernacular expression on political issues is important, that a national 'right to protest' is a valued part of democratic Britain, and that Trafalgar Square is a site in which that expression should be allowed, even if it must be controlled by officials who retain the ability to approve the use of the site. As I have been arguing, the Square's value lies in its role as a place where non-official versions of the nation can be advocated, even if this sometimes ends in violence.

In *The Guardian*, Martin Kettle contextualised the Poll Tax Riots with previous violence in the Square: 'Nelson has looked down from his column

---

99   House of Commons Debate, 2 April 1990 Col 895.
100 *Ibid.*
101 Anonymous, *Poll Tax Riot: 10 hours that shook Trafalgar Square* (London: Acab Press, 1990), 67.
102 John Carvel, 'Riots should not end demonstrations in Whitehall: Imbert cautions against rally ban', *The Guardian* (6 April 1990), 4.

on many Battles of Trafalgar Square over the years. Some have become legends. Many have been forgotten'.[103] However, the use of the Square made such battles memorable, because they formed part of a long history of protests with national visibility. During the Cold War period, these often had an international focus, with groups negotiating a domestic stance on events and decisions in places as distant as South Africa, the US and the Soviet Union. The use of the Square also included royal events in which the Empire had an important symbolic role, even if the way the Empire was understood was ambivalent or even nostalgic. Imperialism was evident in the AAM's rejection of apartheid that was sharpened by an uncomfortable sense of culpability. CND's campaign was also influenced by a relationship with Empire through its role as a 'moral authority' that could help to build peace based on a progressive morality, with roots in nineteenth-century versions of imperialism, as a process of civilising subject peoples.[104] Much like the symbolism in the Square itself, in this period the Empire was there if you chose to see it, including through reframed international relationships or ways of thinking about Britain's place in the world. By the twenty-first century, this had become even more subtle, although still evident, for example, in the coverage of the Fourth Plinth contemporary art scheme. As discussed in the next chapter, just as the Square had been used to build solidarity by showing protest groups their own membership in the twentieth century, events in the twenty-first saw the Square used to show the whole nation a unified vision of itself.

103 Martin Kettle, 'In the kingdom of the blind', *The Guardian* (2 April 1990).
104 Jodi Burkett, 'Re-defining British morality: "Britishness" and the Campaign for Nuclear Disarmament 1958–68', *Twentieth Century British History* 21/2 (2010), 185–186.

Millennium London (2000–2012)

On 6 July 2005, the International Olympic Committee (IOC) announced London's selection as the host city for the 2012 Olympic and Paralympic Games. London's competitors included New York, Moscow and Madrid, although its final rival after several rounds of voting was Paris. In Trafalgar Square, at least 10,000 people had gathered to watch the announcement of the winning host city on large screens at a 'Thank You UK' party, the largest of several official celebrations across the country.[1] The announcement of London's success was met with euphoria in the Square, as athletes, officials, celebrities and the public celebrated. This moment of national delight and pride was expressed in a tone of disbelief in *The Guardian*, which glossed London's win as a British one: 'in these affairs a rhythm has long been established: Britain goes for a big sporting event, Britain's representatives insist they can get it, Britain loses horribly'.[2]

The next day, in the morning rush hour of 7 July 2005, four bombs went off in London. Three exploded on Underground trains near Liverpool Street, Edgeware Road and Kings Cross stations. A fourth exploded on a bus in Tavistock Square, in central London, not far from the British Museum. This was a significant terrorist event even in a city with a history of bomb attacks, and fifty-two people were killed and around 770 injured.[3] When

1   *London 2012*, 'Rachel Stevens and Melanie C join thank you party in Trafalgar Square', press release (4 July 2005) <http://www.london2012.com /news/media-releases/bid-phase/rachel-stevens-and-melanie-c-join-thank-you-party-in-tra.php > accessed 21 Jan 2009.

2   Sean Ingle, 'The greatest comeback of all', 6 July 2005, <http://www.guardian.co.uk/ sport/2005/jul/06/Olympics2012.politics2 > accessed 12 April 2013.

3   '7 July Bombings: Overview', *BBC News* (7 July 2005), <http://news.bbc.co.uk/2/ shared/spl/hi/uk/05/london_blasts/what_happened/html> accessed 23 August 2010.

news of the bombings reached him, London's Mayor, Ken Livingstone, was in Singapore for the Olympics host city announcement. In an emotional and improvised statement, he laid out the message he was to repeat over the following weeks: London's diversity was a source of strength, and its multiculturalism would not be damaged by terrorism.

A narrative of national 'unity in diversity' also shaped the official Olympics events, and Trafalgar Square linked the two through its use as the venue for official public recognition of both occasions. This connection was evident in newspaper reports on 7 July:

> Across London, the sense was the same: a barely comprehensible lurch from limitless jubilation to a very provisional emotion, mixing horror and bafflement in equal measure [...]. Of the celebrations the evening before few signs remained: London's night armies had long removed most of the bunting from Trafalgar Square, along with the boards saying 'Thank You London!' from the foot of Nelson's column.[4]

As a central site for official responses to both events, Trafalgar Square was subject to repeated reference, and helped to frame how Britishness was understood. The period of July 2005 is the first main focus of this chapter. Also during this time, a relatively new programme of temporary sculptural installations was on display on the Square's previously vacant Fourth Plinth. These works, in large part because of their location in the Square, were almost inescapably national in content, and they are analysed in the second half of this chapter.

In 2005, Trafalgar Square had only recently been refurbished. The changes to traffic arrangements, layout and public amenity were part of Livingstone's plan to transform 'London's central square' into, as he put it, a 'unique venue for innovative events'.[5] The 2003 refurbishment of the Square

---

4   Oliver Burkeman, 'From Olympic jubilation to bafflement and horror', *The Guardian* (8 July 2005), <http://www.guardian.co.uk/uk/2005/jul/08/london.politics1/print> accessed 27 Jan 2009.

5   *London 2012*, 'Trafalgar Square Festival makes a big splash', press release (5 August 2005), <http://www.london2012.com/news/archive/post-hid-2005/trafalgar-square-festival-makes-a-big-splash.php> accessed 4 March 2009. For more on the changes to the Square in 2003 see Williams, *Anxious City*.

emphasised accessibility and openness, highlighting its role as a gathering place. In addition, by promoting a strong multicultural programme of events for the Square, the GLA constructed a physical and metaphorical space in which it intended to represent an official metropolitan govern-ment vision of twenty-first century London. Through the GLA's official support, London's multiculturalism was repeatedly highlighted in Trafalgar Square through its use for events to promote the positive aspects of cul-tural diversity. This was a purposeful strategy on the part of the GLA and Mayor Livingstone. In 2005, for example, Livingstone launched 'Eid in the Square', an annual celebration of the Muslim religious holiday marking the end of Ramadan. Since then religious and cultural festivals, such as Eid, the Hindu festival of light, Diwali, Chinese New Year and Hanukkah have all been celebrated at the site.

Some aspects of Trafalgar Square's history have been germane to this project of reinventing the Square as a more 'multicultural' place. The Square's longstanding use as a gathering place to express public sentiment about national issues, for example, is a tradition the GLA has chosen to support. Although explicit highlighting of the symbols of Empire and naval power did not figure strongly in the redesign of the Square, the past that the Square's statues and monuments represent are part of what has made London into the super-diverse 'world in one city' that Livingstone championed as mayor.[6] This is because many of London's non-Anglo inhab-itants have family roots in the countries that comprised parts of Britain's former empire in Asia, the West Indies and Africa. In a debate on the future of London, Massey called this the 'external geography' of the city, 'the connections that run out from "here": the trade routes, investments, political and cultural influences; power relations of all sorts run out from here around the globe and link the fate of other places to what is done in London'.[7] In this way, the imperial history represented in Trafalgar Square has shaped its present and future through the 'external geography' evident

6   'London: a world in one city', *The Guardian* (21 January 2005), <http://www.guard ian.co.uk/britain/london/0,,1394802,00.html> accessed 14 November 2007.

7   Doreen Massey, *World City* (London: Polity, 2007).

in the multiethnic makeup of London's population. Artist Yinka Shonibare MBE, discussed below, engaged with this aspect on London directly with his Fourth Plinth installation in 2010–2011.

The analysis of these events in this chapter is framed by a brief discussion of modern British multiculturalism. It has been one of the notable aspects of the GLA's vision for the Square since its refurbishment in 2003, and was also central to the narrative surrounding both the Olympics bid win and the London bombings. More generally, diversity was a common theme of Fourth Plinth installations and media reactions to them, and so warrants a brief discussion.

Britishness, London and multiculturalism

The early 2000s saw renewed public discussion over official Britishness. An example, largely championed by previous Prime Minister Gordon Brown, was the 'Life in the UK' Test, an exam on British values for applicants for British citizenship that was introduced in November 2005. Passing this exam became a requirement for all people wanting to acquire UK citizenship or permanent residency, and demonstrates the attempt to codify aspects of Britishness in reaction to immigration. Other related suggestions of 'Britain Day', based in part on the example of Australia Day, and 'citizen packs' for British residents turning eighteen, that explain what the state expects from adults in the UK, were not adopted.[8] These proposals do, however, illustrate official concern about British culture, diversity and the relationship between the individual and the state, and occurred in a context of perceived challenges to national identity. These included European enlargement and the arrival of new groups of mainly eastern Europeans who moved to the UK for work; ongoing devolution movements

---

8   'Ministers proposing "Britain Day"', *BBC News* (5 June 2007) <http://news.bbc. co.uk/2/hi/uk_news/6721239.stm> accessed 19 November 2007.

in the 'Celtic fringe', especially Scotland, which began to gain real political momentum in 1997; and increased political concern about the potential for terrorism.[9]

In terms of the inclusion of Muslims and other minority racial or religious groups into the British imagined community, London was notably unique. In 2008, about one million Londoners were Muslim, representing approximately two-thirds of the Muslim population of Britain. Around forty-two per cent of London's population was non-white, as opposed to eight per cent in the UK overall, and a third of London residents were born outside the UK, compared with twelve per cent of the overall population.[10] In addition, London accommodated Britain's largest increase in numbers of immigrants, with a forty-four per cent increase in the number of people in London who were born abroad between 1991 and 2001. Although this measure also captures white immigrants, many of whom may have come from the former Commonwealth countries or Europe, among the leading five places of birth outside the UK were India, Pakistan and the Caribbean.[11] These statistics give a sense of the makeup of London's 'visible minorities'.[12]

The visibility of these populations contributed to an ongoing public debate about the value of diversity, and after the 7 July 2005 bombings, existing disquiet about British multiculturalism and immigration intensified. In August 2005, for example, Shadow Home Secretary, David Davis, called for the Labour Government to 'scrap its outdated policy of multiculturalism' and move towards a society in which immigrants were more integrated.[13] The reaction to Davis' comments was reflected in a BBC News

9   Yasmin Alibhai-Brown, 'Muddled leaders and the Future of British National Identity', *The Political Quarterly* 71/1 (2000), 26–30. See also Yasmin Alibhai-Brown, *Imagining the New Britain* (New York: Routledge, 2001).

10 National Statistics Online, 'London: population and migration', 8 June 2010, <http://www.statistics.gov.uk/cci/nugget.asp?id=2235> accessed 23 August 2010.

11 'Born Abroad: An Immigration Map of Britain. Countries of Birth', *BBC News* <http://news.bbc.co.uk/2/shared/spl/hi/uk/05/born_abroad/countries/html/overview.stm?ia_percent01_des#table_1a> accessed 13 September 2010.

12 See Brown, *Imagining the New Britain*.

13 'Davis attacks UK multiculturalism', *BBC News* (3 August 2005), <http://news.bbc.co.uk/1/hi/uk_politics/4740633.stm> accessed 4 March 2009.

online discussion in which most respondents were broadly supportive of
his approach, as in this example:

> Multiculturalism is a myth. If you don't enforce integration into the country's society
> you end up with what Britain has today, microcosm societies living side by side with
> no interaction and no understanding of each other. It is time to end all this idiotic
> political correctness and to stand up for British culture.[14]

For this member of the public, the main issue seemed to concern the boundaries of Britishness and the necessity for new-comers to adopt its existing
narratives. Again focusing on the issue of integration and the parameters of
the national discourse, Prime Minister Gordon Brown stressed the importance of 'the balance between diversity and integration'.[15] While invoking a
flexible and changing discourse of national identity able to include a diverse
and multicultural population, Brown insisted on cultural boundaries that
he expressed as very general values: 'liberty for all, responsibility by all and
fairness to all'. Despite the generic terms he used, this statement suggested
that he understood Britishness as a bounded discourse, the limits of which
were being tested by the presence of multicultural communities. For Brown,
the flexibility of national identity was finite, and the UK risked serious
social division if it did not address that question of 'how diverse cultures,
which inevitably contain differences, can find the essential common purpose without which no society can flourish'.[16] Here he described a subtle
threat to national unity and division, identifying cultural diversity as the
source of this potential conflict.

Academics and journalists also took up the question of the flexibility
of national identity, with Yasmin Alibhai-Brown, for example, positioning multiculturalism at its core. She argued that Britishness is not merely
enhanced by recognising its potential for racial inclusion, but instead is an
identity in which blackness is inherent, stressing that 'there is no modern

---

14   *Ibid.*
15   Gordon Brown, 'The Future of Britishness'. Speech to the Fabian Future of Britishness
     conference (14 January 2006) <http://www.fabian-society.org.uk/press_office/
     news_latest_all.asp? pressid=520> accessed 21 May 2007.
16   *Ibid.*

British identity without Indian food, black music, Salman Rushdie and [black television presenter] Trevor Phillips.[17] In a book published only a year before the Olympics win and the London bombings, Paul Ward also argued for the centrality of race in current and future versions of Britishness, as well as the need to jettison old attitudes about the inferiority of black and Asian people:

> Most white Britons have, after half a century of black and Asian immigration and descendency, begun to come to terms with the need to define a new way of being British, because black and Asian Britons have insisted that they do so. Many, probably most, British people, white and black, have seen the experience of immigration and ethnic diversity as a positive benefit to British culture and identity.[18]

In these accounts, one of the issues driving the discourse of Britishness is multiculturalism and the question of its limits. In light of this ongoing contest over the value, impact and extent of multiculturalism, official statements about the Olympics bid and bombing vigil that emphasised positive diversity were both highly political and important to debates about Britishness.

Olympics bid win

Official coverage of London's successful bid to host the 2012 Olympics focused on national jubilation, and media coverage of the event was similarly characterised by descriptions of popular elation when London's win was announced. On the Number 10 website, Prime Minister Tony Blair commented: 'I am just overwhelmed. I can scarcely believe it actually, it

17 Alibhai-Brown, 'Muddled leaders', 29.
18 Ward, *Britishness*, 140. There is an enormous body of literature on British multiculturalism, but some useful sources include Paul Gilroy, *After Empire: Melancholia or Convivial Culture* (London: Routledge, 2004); and Tariq Modood, *Multiculturalism: a Civic Idea* (Cambridge: Polity Press, 2013).

is just the most fantastic thing. It is extraordinary, momentous, a great honour and a privilege, and I am just really proud of our country today.'[19] However, the delight over the Olympics win was not universal, with some people expressing disinterest. Directly after the win, public attention began to focus on the enormous budget for the development of infrastructure and the opportunity cost of spending so much on the Olympic Games. Although these did not reflect the majority view represented in the mainstream media, this ambivalence to the Olympic Games hints at the range of public responses to the win; some people were highly critical and questioned the narrative of national success that accompanied official reactions.

Prior to the announcement, London's diversity and multiculturalism was a major theme of both the official narrative around the event and London's bid, and was reportedly one of London's most attractive qualities for the IOC members who considered potential host cities. The Muslim Council of Britain (MCB), a body representing over 400 Muslim community groups across the UK, was one organisation that provided the official multicultural support central to the Games bid. According to the MCB's Secretary General, Iqbal Sacranie, the MCB fully supported the Games' multicultural vision: 'We [the MCB] want to play an active role in welcoming visitors and ensuring the London Games are fully reflective of our multicultural, multi-faith society'.[20] Lord Coe, Chairman of the London 2012 bid committee, presented the MCB's support as an example of the multiculturalism at the heart of London's bid, claiming it 'demonstrates our vibrancy as a truly diverse city where every athlete and visitor can feel at home'.[21] The newspaper coverage of the bid also reflected this aspect of London: 'Those making the final decision [on the Olympic host city] were impressed, apparently, with our openness to other countries and cultures...

---

19   Tony Blair, 'PM's comments on 2012 Olympic Games', 6 July 2005, <http://www. number10.gov.uk/Page7832> accessed 13 September 2010.

20   *London 2012*, 'Muslim Council of Britain lends support to the London Olympic Bid', press release, 23 August 2005 <http://www.london2012.com/ news/media-releases/post bid 2005/muslim-council-of-britain-lends-support-to-the-london-ol. php?stylesheet=normal> accessed 4 March 2009.

21   *Ibid.*

[London] is a global centre for transport, money, communications and migration, with its 50 separate ethnic communities and its more than 300 language groups'.[22]

In Trafalgar Square itself, the multiculturalism that was a part of the official and media representation of London's character also informed coverage of the 'Thank You UK' Olympics bid party. People in the Square at the celebration party clearly delighted in media photographs, enjoyed a range of musical performances and other entertainment: 'in keeping with London's cultural diversity, which has been one of the bid's selling points, there will also be a carnival group and Indian drummers [at the party]'.[23] Black and Asian athletes, such as Dame Kelly Holmes, were high-profile participants in the celebrations. London's multiculturalism was an important aspect for the 2012 Olympics bid and a focus of official preparations for the celebrations. The emphasis on diversity within the bid itself and in general reactions to the win is prominent in official and media sources, suggesting that this was an important part of what authorities wanted to highlight about London. Other accounts, such as newspaper descriptions of the event in the Square, do not focus as strongly on multiculturalism or diversity as a significant aspect of the celebrations. This complicates the centrality of racial and ethnic diversity that was the centerpiece of the bid. This could be because London is unrepresentative in the extent and range of its diversity compared with the rest of the UK. It could also be that diversity is so commonplace in London that it was almost not worth mentioning outside the official discourse. It is certainly the case that the official 'top down' treatment of the event did not coincide with more popular descriptions.

An additional narrative, portrayed in the press and by political leaders, described the Olympics bid as a win for all of Britain, even though it

22   Jackie Ashley, 'The world in one city', *The Guardian* (7 July 2005), <http://www.guardian.co.uk/politics/2005/jul/07/london.Olympics2012/print> accessed 27 January 2009.

23   Hugh Muir, 'Country gears up to celebrate – or commiserate – in style', *The Guardian* (6 July 2005), <http://www.guardian.co.uk/uk/2005/jul/06/olympics2012.communities2> accessed 27 January 2009.

was centred on London. For example, in the House of Commons on the day the winning bid was announced, Foreign Secretary Jack Straw spoke about his own feelings, as well as those of the wider national community:

> The nation has united behind this vision. The latest polls show that 80 per cent of people backed the bid and more that 3 million people sent individual pledges of support [...]. Securing the games is one of the greatest international prizes for any nation. Like every hon. Member, I have always been proud of my country but today I am prouder than ever.[24]

Here, London's win was a British win, an event for the whole nation to celebrate. Straw appears to conflate London with Britain, and similarly, in this example from *The Guardian*, the Games bid was primarily London's victory:

> Brilliant for London as a whole, for which the Olympics will provide a thrilling validation and climax to its 21st-century re-emergence as an open, multiracial and dynamic world city. And brilliant too, we must ensure, for other cities and other parts of Britain too. Many of them will play a role in 2012 and their interests must not be forgotten, even if in the end the focus inevitably concentrates on this extraordinary and wonderfully diverse capital city of ours.[25]

Other voices, however, insisted that while London was sometimes symbolic of the nation, it was not completely representative of it. After the Olympics bid, there were several pleas in the House of Commons, like that of the Member for the Northern Irish constituency of North Antrim, Rev. Ian Paisley, for areas outside London to benefit from the Olympics, and Foreign Secretary Straw was careful to underscore the national benefit of the Games: 'they are not just a games for London. They will leave a legacy for the entire country. Olympic competitions will be held in Glasgow, Cardiff, Weymouth, Birmingham, Manchester and Newcastle'.[26]

24   House of Commons Debate, 6 July 2005 Vol 436 Col 404–405.
25   *The Guardian*, 'A famous victory', *The Guardian* (7 July 2005) <http://www.guardian. co.uk/society/2005/jul/07/communities.olympics2012> accessed 4 March 2009.
26   House of Commons, 6 July 2005, Col 405–410.

As with the issue of multiculturalism, the reporting of the party in the Square did not focus explicitly on this issue. However, the use of Trafalgar Square for what was arguably a London-specific occasion helped to conflate London and Britain, in part because Trafalgar Square is commonly understood as a national site. Many of London's prominent sites have national significance because of their functions, such as Buckingham Palace or the Houses of Parliament, but for Trafalgar Square, its national significance in this instance rested on its historical symbolism.

For example, a prominent theme in reports of the Olympics bid used Trafalgar Square and Nelson's Column to construct a national narrative based on general victory over the French that referenced the 1805 Battle of Trafalgar. The media emphasis on French inferiority hinged on London having beaten Paris by a close margin of fifty-four IOC votes to fifty. It also occurred in the political context of a G8 summit meeting including the French and British governments, before which French President Jacques Chirac had insulted British cuisine, joking that 'mad cow disease [was...] the only British contribution to European agriculture. "You can't trust people who cook as badly as that," [Chirac] added'.[27]

The G8 summit political tensions were linked to a much longer history of conflict between the UK and France, with the title of a *Guardian* editorial, 'A famous victory', alluding to both the Battle of Trafalgar and the Olympics win. The editors began by describing the scene in Trafalgar Square when the outcome of the Olympics competition was announced: 'On the ground, there was instant celebration too. In Trafalgar Square, there were hugs and kisses for the cameras in front of the banner that simply said Thank You.'[28] The article then emphasised the 'political frissons' of London's win given the state of British–French relations. This reminded the reader that Trafalgar Square, with its historical representation of British victory over the French at Trafalgar, also framed modern national celebrations over the same adversary. Other newspaper coverage of the Olympics bid

---

27   Henry Porter, 'For that week in July, I salute Blair', *The Guardian* (18 December 2005) <http://www.guardian.co.uk/politics/2005/dec/18/labour> accessed 4 March 2009.
28   'A famous victory', *The Guardian* (7 July 2005).

celebrations also identified British success with French defeat, linking it with the symbolism of Trafalgar Square. In one example, the reporter used the celebrations in the Square to link the Olympics win to both VE Day and the Battle of Trafalgar: 'Yesterday, in Trafalgar Square, amid scenes the likes of which Londoners had not seen since VE Day and, probably, since Nelson's victory over the French in 1805 [...]'.[29] The reference to VE Day invoked the popular narrative of the wild celebrations in the Square in 1945. *The Sun* further emphasised the links between the Battle of Trafalgar and London's victory over Paris against a background of general British superiority over the French:

> More than 10,000 ecstatic Brits celebrated yesterday's triumph over the French – in Trafalgar Square [...]. The occasion could not have been better scripted – in this the 200th anniversary of our greatest naval success. Admiral Lord Nelson looked down from his column at the historic events unfolding below him [...]. Horatio Nelson himself could not have been prouder. The French had been sunk again.[30]

By referring to the structures of the Square, this report took advantage of the symbolism of the site, using aspects of the national past to frame the Olympics bid success with a narrative of longstanding competition between the British and the French. The details of the Battle of Trafalgar were less important than the fact of previous victory, and as with previous uses, the specific environment of the Square served as a means to interpret the contemporary events there. Not everyone, however, was comfortable with this approach. Jackie Ashley, for example, warned against interpreting London's success as 'a triumph for the British way, a vindication of Britain's approach to Europe and a final trouncing of the French'.[31] But even though

---

29 Mark Honigsbaum, 'Patriotism and pop mark victory celebrations', *The Guardian* (7 July 2005) <http://www.guardian.co.uk/uk/2005/jul/07/olympics2012.olympicgames> accessed 27 January 2009.

30 V. Wheeler and J. Blair, 'Nelson looked down as Britain cheered victory at Trafalgar; London 2012: 10,000 go wild', *The Sun* (7 July 2005), 4. For more on the role of the French in coalescing Britishness, see Linda Colley, *Britons: Forging the Nation, 1707–1837* (London: Pimlico, 1992).

31 Ashley, 'The World in One City', 7 July 2005.

she was careful to play down any prejudice, Ashley's caution against it serves to highlight the more general anti-French bias in the narrative of British national identity in coverage of the celebrations.

These examples demonstrate the divergence between official narratives of the Olympics bid and popular ones of the event, especially in terms of multiculturalism and diversity. Powerful officials in the London and national governments who strongly supported the Olympic bid team, used events in the Square to promote messages about identity that did not appear to be central aspects of the event in media reporting on the use of the Square. As with the VE Day celebrations in 1945, during which individual accounts rarely mentioned the national and imperial symbolism that official and media reports highlighted, the official emphasis on multiculturalism in the 2005 Olympics bid was not common to more popular accounts.

London United

The morning after the Olympics bid party in the Square, four bombs exploded on London's public transport system, killing fifty-two people. London's Mayor Ken Livingstone's first statement after the attacks indicated what city officials hoped the public response would be. He outlined the London authorities' version of the appropriate response to the bombings and set the tone for the official narrative of the attacks:

> [The bombers] seek to divide Londoners. They seek to turn Londoners against each other...[but] the city of London is the greatest in the world, because everybody lives side by side in harmony. Londoners will not be divided by this cowardly attack. They will stand together in solidarity alongside those who have been bereaved [...][32]

32   Tim Dowling, 'Fear in the city', *The Guardian* (26 July 2005) <http://www.guard-ian.co.uk/uk/2005/jul/26/july7.uksecurity3/print> accessed 9 February 2009.

This narrative was reinforced at an officially-sponsored vigil in Trafalgar Square a week later that commemorated the attack's victims. Livingstone's office flagged the official response in a press release two days before, which clearly set out how the London government wanted the public to understand London in the context of the bombings:

> At 6pm Londoners are invited to a vigil in Trafalgar Square to remember those who died, to show that London will not be moved from our goal of building an open, tolerant, multi-racial and multi-cultural society showing the world its future [...].[33]

Fifty thousand people attended the vigil on the evening of 14 July, and many more observed a two-minute silence in London and across the UK at midday.[34] As with some descriptions of the 1945 celebrations, media reports of the event included reference to the collective experience of attending the vigil by mentioning the number of people who could gather at the site: 'it was standing room only in Trafalgar Square [...] as thousands of people shouldered their way in to attend a vigil'.[35] Furthermore, it was unusual, according to *The Guardian*, for Londoners to gather to hear such positive messages about themselves: 'last night in Trafalgar Square was a novel experience. A huge crowd, standing in their thousands in blinding sunshine, to be told again and again that they live in one of the greatest cities in the world'.[36] As with the 'crowd-feeling' in 1945, the event was characterised in terms of the large number of people in the Square that demonstrated the significance of the event, as well as the emotional impact

33   Greater London Authority, 'London united in defiance of terrorist attacks', press release (12 July 2005) <http://www.london.gov.uk/ view_press_release.jsp? releaseid=5325> accessed 4 March 2009.

34   Greater London Authority, 'Trafalgar Square vigil, 14 July', press release (14 July 2005) <http://www.webarchive.org.uk/pan/12334/20050729/www.london.gov.uk/mayor/london-bombings/traf-sq-vigil.html> accessed 9 Feb 2009.

35   Emma Griffiths, 'London unites for healing vigil', *BBC News* (14 July 2005) <http://news.bbc.co.uk/1/hi/ england/london/4684453.stm> accessed 13 February 2009.

36   Jonathan Freedland, 'The world in one city', *The Guardian* (15 July 2005) <http://www.guardian.co.uk/uk/2005/jul/15/july7.uksecurity9/print> accessed 9 February 2009.

of the crowd itself. Other reports emphasised the Square's symbolic value and historical role as a gathering place by contrasting the bombing vigil's two-minute silence with previous uses: 'At Trafalgar Square, which so often has served as the meeting quarters for Londoners to protest and to pray, to sing and to celebrate, people began to gather well before noon [for the two-minute silence]'.[37] In this example, the Square's importance as a national site is reinforced through its description as a gathering place for people to participate in significant national moments.

The event was framed by the prominently displayed slogan 'London United', a play on words that suggested a football team name as well as metropolitan solidarity. This set the tone for the speakers' messages of tolerance, togetherness and the value of diversity. Livingstone was the first speaker, and in his emotional speech he pleaded with Londoners to resist accusation and division as a result of the bombings:

> I have watched this city transform in my lifetime as a beacon of what the world can be and I hope and pray will be. It is a city that embraces change. It is a city which is the most tolerant in the world.

> Those who came here to kill last Thursday had many goals, but one was that we should turn on each other like animals trapped in a cage. And they failed. They failed totally and utterly. There may be places in the world where still that would have happened, but not here.[38]

During the vigil, media coverage singled out views from participants that echoed the tone of unified resilience. For example, in *The Guardian*, 'Stuart Giddens, 26, of Hatfield, Hertfordshire' said he attended the vigil because it was 'really important that we show that we won't be alarmed by anything that these people can throw at us [...]. There are people here from every walk of life, all different types of people and the point is to show that we

---

37 Todd Richissin, 'For two long minutes, London quietly remembers', *LA Times* (15 July 2005) <http://www.latimes.com/news/nationworld/wire/ balte.britain15jul15,0, 3681895.story> accessed 9 Feb 2009.

38 Ken Livingstone, 'London United', 14 July 2005 <http://www.youtube.com/ watch?v=6BSIBPsbL9c> accessed 4 March 2009.

are stronger than them, collectively'.[39] Other reports noted not only this spirit of collectivity, but also how different the participants were from each other in terms of social background: 'Some [...] were clearly tourists who had got caught up in events while visiting the square. Cyclists and students lined up alongside men in suits and others in bandanas along the square's steps, fountains and walls'.[40] Here, the message of 'London United' that was a strong theme of the official message broadcast from Trafalgar Square also appeared in descriptions of participants, and the Square functioned almost as a *tableau vivant* for the modern London that officials such as Livingstone and Olympics bid chairman Lord Coe promoted: diverse, multicultural, unified, tolerant and resilient.[41]

Furthermore, at the vigil Coe reinforced the connection between the ongoing development of Olympics facilities and the bombings: 'As we move forward, we will never, ever forget those who suffered so grievously last Thursday, but our efforts and actions over the next seven years are dedicated to them'.[42] In the speeches at the vigil, the Olympic Games were shorthand for a particular vision of London as positive, successful, multicultural and tolerant, as the diversity of London that had been a selling point in the Olympics bid was revised in official attempts by the media and London authorities to knit the city together after the bombings. By linking the two events, Olympic success was used in response to the bomb attacks to remind audiences of the desirability of London's multiculturalism. Closs Stephens points out that 'the Olympics narrative was mainly deployed to affirm London – and through it Britain – as a multicultural, multiethnic

39   'London pays tribute to blast victims', *The Guardian* (July 14 2005) <http://www.guardian.co.uk/uk/2005/jul/14/july7.uksecurity12 > accessed 9 Feb 2009.
40   Griffiths, 'London Unites', *BBC News*.
41   For more on the role of commemorative activity in narrating national identity, see John R. Gillis, *Commemorations: The Politics of National Identity* (Princeton: Princeton University Press, 1994).
42   'Thousands attend Trafalgar Square vigil', *The Telegraph* (11 July 2005) <http.//www.telegraph.co.uk/news/1494022/Thousands-attend-Trafalgar-Square-vigil.html> accessed 20 September 2010.

community' in the official reaction to the London bombings.[43] The theme of the value of diversity, and the importance of national unity, was repeated in the House of Commons on the day of the vigil, and this exchange also hinted at the reasons behind the message of positive multiculturalism that had been adopted by officials as they sought to shape public opinion:

> Dr Ashok Kumar [Middlesbrough, South and East Cleveland Lab]: Following the tragic and evil events of last week caused by terrorism, there is already fallout in the Asian community – not just the Muslim community but the Hindu community and the Sikh community – to which I belong. Given the serious effects on multiculturalism and the great society that we have built up... it is important that we... show strong support for the spirit of multiculturalism[44]

The reference to a 'backlash' against Asians suggests that the official message of 'London United' was intended to counter perceived racial prejudice in the larger population. For example, on the day of the bombings, according to a report in *The Independent*:

> The explosions prompted the Islamic Human Rights Commission [...] to issue the extraordinary advice yesterday that no Muslim should travel or go out unless strictly necessary, for fear of reprisals. The Muslim Association of Britain said women in headscarves were at particular risk, asked police to consider extra protection for mosques and Islamic schools, and also warned Muslims against unnecessary journeys.[45]

Furthermore, in an interview with the BBC a year after the bombings, Ken Livingstone hinted that London officials had been aware of the potential for racially motivated conflict after the London bombings. He also identified how he had planned to pre-empt such conflict:

43 Angharad Closs Stephens, '"Seven Million Londoners, One London": National and Urban Ideas of Community in the Aftermath of the 7 July 2005 Bombings in London', *Alternatives* 32 (2007), 155–176.

44 House of Commons Debate, 14 July 2005 Vol 436 Col 969.

45 Ian Herbert, 'Muslims told not to travel as retaliation fears grow', *Independent* (8 July 2005) <http://www.independent.co.uk/news/uk/this-britain/muslims-told-not-to-travel-as-retaliation-fears-grow-497998.html> accessed 20 September 2010.

KEN LIVINGSTONE: The only thing, the only doubt [about London's emergency planning] was, would it [a terrorist attack] unleash tensions between Londoners. And therefore everything I had planned to say on that had to be directed at making sure we got through this, that London was united

INTERVIEWER: Were you surprised that London [...] became the target [...] of homegrown Muslims?

KEN LIVINGSTONE: Nothing was a surprise, except that Londoners stood united. We assumed that there would be some tensions, some thugs would go out and beat up the first Asian-looking person they found and so on and I think that a lot of London's ethnic minorities wondered when this does happen won't people turn on us and that they didn't I think has left London immeasurably stronger.[46]

These examples help explain why the theme of unity and the positive value of multiculturalism was consistently repeated by London officials. Livingstone's apparent surprise that 'Londoners stood united' suggests that officials viewed London's social cohesion as fragile, rather than the open, tolerant and robust selling-point of the Olympics bid. That the mayor of London should express any doubt about the success of London's multiculturalism aligned with the ongoing debate about the success and value of British multiculturalism and immigration more generally. One survey of contemporaneous attitudes also appeared to question the official emphasis on unity. Results of a poll by ICM research found that: 'Nearly two-thirds of Muslims [...] had thought about their future in Britain after the [London bomb] attacks, with 63% saying they had considered whether they wanted to remain in the UK'.[47]

Disagreement over multiculturalism underscores the value to officials in hosting the bombing vigil in Trafalgar Square, because the visible presence in the Square of a crowd apparently committed to London's diversity

---

46   'Ken Livingstone – The week that changed London', *BBC News* (7 July 2006) <http://www.bbc.co.uk/london/content/articles/2006/07/06/7july_oneyearon_videos.shtml> accessed 20 September 2010.

47   Vikram Dodd, 'Two-thirds of Muslims consider leaving UK', *The Guardian* (26 July 2005) <http://www.guardian.co.uk/uk/2005/jul/26/polls.july7> accessed 10 May 2010.

helped to demonstrate the value of a multiethnic narrative of Britishness. By using the Square to promote a version of London that was multicultural and peaceful, authorities promoted a type of city and nation they wanted to encourage, imagining it as a particular type of national community. While this use of the Square was not directly contested within the space itself, the examples above show that the larger message of multicultural tolerance was not universally accepted. The bombings fuelled an ongoing debate about the relationship between multiculturalism and 'Britishness', which included the view that multiculturalism was seen to have allowed, and possibly even encouraged the bombers. In light of this, the bombing vigil was a political statement about London's official stance towards multiculturalism: that it made London unique, powerful, attractive and dynamic; that it had gained the city (and the nation) the Olympics; and that it would help the city's residents to come together to support each other after the bombings. In using Trafalgar Square as the venue for the official bombing vigil, London officials attempted to leverage the positive and future-oriented vision of diversity that had been emphasised in the Olympics bid celebrations to reassure Londoners and help maintain order following the bombings.

Media reporting also constructed the Trafalgar Square vigil in terms of its relationship with aspects of the national past. For example, some speeches made at the vigil alluded to London's experience of war, such as the section of poet Ben Okri's speech that said London had grown stronger and more beautiful following past 'bombings, burnings and wars'.[48] Other reports highlighted the mood of unity and resilience at the vigil itself, subtly linking it to the view that Londoners showed the same stoicism during World War Two.[49]

The reporting of the two-minute silence and the Trafalgar Square vigil in the mainstream newspapers, including *The Guardian*, *The Daily Telegraph*, *The Times*, *The Daily Mail*, *The Sun*, and *The Evening Standard*,

---

48 'Vigil for London bombing victims' *BBC News* (14 July 2005) <http://news.bbc. co.uk/1/hi/uk/4684207.stm> accessed 9 February 2009.

49 See Richissin, 'For two long minutes' and Griffiths, 'London unites'.

uniformly adhered to the themes of unity, dignity and calm, with many using the Blitz as a historical reference point to describe Londoners' response to the bombings. *The Sun* used the theme of the Blitz and World War Two to encourage participation in the two minute silence, thus helping to transmit official messages regarding the 'appropriate' public reaction to the bombings.[50] Manthorpe argued that the 'rhetoric of the Blitz was pervasive' in the newspaper reporting the day after the bombings, quoting the leader article in *The Sun*: 'Our spirit will never be broken: Adolf Hitler's Blitz and his doodlebug rickets never once broke London's spirit'.[51]

Beyond the event in the Square, the emphasis on calm resilience permeated reporting on the bombings, with reports of a dignified and compassionate public response to the attacks used to link the public response to the 2005 bombings with 1940–1941: 'The spirit of the Blitz was invoked shortly after the bombings of Thursday July 7, and it seemed to resonate immediately. Those directly affected by the attacks [...] did indeed behave with courageous stoicism, and Londoners took a little reflected pride in their dignity'.[52] However, despite mention of Londoners' stoic response to the bombings, there were other reports of fearfulness and caution. One reporter summed up this feeling: 'I must admit I'm now afraid; afraid that another attack is imminent, afraid of the idea of 3,000 armed police on the streets, afraid that London will never be quite the same again'.[53] A more complex version of public reaction to the bombings emerges from these examples, one that does not conform to the message of resilience and calm that was emphasised in official and media narratives, but constructs London as vulnerable and anxiety-inducing.

Again as with 1940–1941, the 2005 bombings were constructed as an attack on the whole of Britain, and emphasised London as a symbol of Britain. For example, Prime Minister Tony Blair conflated London and Britain in describing Londoners' reactions to bombings:

50   R. White, 'Make stand today', *The Sun* (14 July 2005), 6.
51   Rowland Manthorpe, 'Spirit of the Brits', *The Guardian* (1 July 2006). <http://www.guardian.co.uk/books/2006/jul/01/featuresreviews.guardianreview29> accessed 30 August 2010.
52   Dowling, 'Fear in the City'.
53   *Ibid.*

> Yesterday we celebrated the heroism of world war two, including the civilian heroes of London's blitz. Today, what a different city London is – a city of many cultures, faiths and races, hardly recognisable from the London of 1945. It is so different and yet, in the face of this attack, there is something wonderfully familiar in the confident spirit that moves throughout the city, enabling it to take the blow but still not flinch from reasserting its will to triumph over adversity. Britain may be different today, but the coming together and the character are still the same.[54]

Blair's description of London 'taking the blow' was very close to Churchill's famous wartime remark that 'London can take it'. This was a commonly repeated media theme as the past was used to frame the present during the bombing commemoration. At the vigil in Trafalgar Square, for example, 'Applause rang out at every mention of London's resilience. It survived the Blitz, it would survive and defy the suicide bombers.'[55] At this event, Trafalgar Square framed the demonstration of what reporters described as 'British' qualities anchored in a historical narrative, which included calm, resilience and unity in the face of attack.

When Blair identified the 'confident spirit' of the city as a British quality, rather than one specific to London, he effectively described London's unique diversity as a national characteristic. The conflation of London and Britain is also reflected in MOA responses that identify London's diversity as a problem for the entire nation. Closs Stephens argues that some of the conflict over London's multiculturalism, as expressed in the national discourse following the bombings, rests with the use of the narrative of the Blitz:

> By invoking the Blitz [...] the history of London is recounted as the history of Britain. By tying the people living in London today into a direct relationship with those who lived in London at the time of World War II, this linear national narrative produces a particular idea of British culture [...] the image of a distinctively white, wartime Englishness.[56]

---

54 House of Commons Debate, 11 July 2005 Vol 436 Part 31 Col 567.
55 Robert Mendick, 'Mayor's Olympic pledge to London bomb survivors', *The Evening Standard* (15 July 2005) < http://www.questia.com/library/1G1–134073398/mayor-s-olympic-pledge-to-london-bomb-survivors-best> accessed 13 April 2013.
56 Closs Stephens, 'Seven Million Londoners', 159–160.

According to this interpretation, by using London as shorthand for Britain, Blair activated a white, English narrative of unity that existed uneasily alongside other messages stressing London's valuable multiculturalism. This was especially problematic given that London's multicultural diversity was a prominent feature of other aspects of the official narrative following the bomb attacks, such as emphasis on the winning Olympics bid. While for the Olympics bid, the Napoleonic wars and victory over the French were historical references, for the bombing vigil, the dignity and resilience of Londoners featured, with particular reference to World War Two and the London Blitz. Both these approaches used the Square to link Londoners in 2005 to the past, but it was an uneasy relationship, complicated by the diversity of contemporary British society.

The Fourth Plinth

In the first decade of the new century, explorations of multiculturalism, diversity and the ethnic makeup of Britain was also a common theme of events and artworks in the Square. This aspect of Britishness in the Square evolved throughout the twentieth century, and can be traced through depictions of Empire and imperial troops in the first half of the century, to conflict over decolonialisation and South Africa from the 1960s, to the reframing of multiculturalism as a virtue by the beginning of the century. Even the presence of Scottish football fans celebrating wins over England, as discussed in Chapter Four, points to debates over Britishness, spatially manifested.

However, the material representations of imperial heroes in the Square have not remained relevant for modern Londoners, and to contemporary viewers its monuments can appear to belong to another age. In a 'super-diverse' city such as London, statues of imperial and martial heroes are not only irrelevant, but are potentially problematic insertions from a national past that often celebrated the control or dominance of other groups. In some cases, the symbolic structures become so familiar that they can almost disappear into the urban background as the meanings ascribed to them by previous generations dissipate.[57]

Ken Livingstone has suggested this is the case for the statues in Trafalgar Square. In 2000, before a major refurbishment of the space, he admitted that 'I think that the people on the plinths in the main square in our capital city should be identifiable to the generality of the population. I have not a clue who two of the generals there are or what they did'.[58] A process of lost meaning, or forgetting, has occurred as the role of the imperial figures in the Square, and what their activities might have represented about the nation and the Empire, fade from public memory: as Darwent says, 'for all its marble and bronze, Trafalgar Square is an homage to forgetting'.[59]

However, even as knowledge of specific names and deeds has faded, the representations of the national past in Trafalgar Square continue to condition the space around them, and such 'forgetting', to the extent that it happens, is not uniform and does not lessen the relevance of the Square for revealing and generating narratives of national identity. Even if most visitors to the Square have not a clue as to who its statues represent, those representations still play an important role in contributing to narratives of contemporary national identity. In part this is because more modern interventions into the space cannot avoid a relationship with the existing monuments in the Square, as well as the Square itself.

One example of such modern interventions is the program of temporary art installations on the Fourth Plinth. This corner of the Square

---

57   See Deborah Cherry, 'Statues in the Square: Hauntings at the Heart of Empire', *Art History* 29/4, (September 2006), 660–697 and Malcolm Miles, '*One & Other*: a picture of the nation in a period of cosmopolitanism?', *The Journal of Architecture* 16/3 (2011), 347–363.

58   Cited in Paul Kelso, 'Mayor attacks generals in the battle of Trafalgar Square', *The Guardian* (20 October 2000) <http://www.guardian.co.uk/uk/2000/oct/20/london.politicalnews> accessed 23 Nov 2011.

59   Charles Darwent, 'Fourth Plinth, National Gallery, London', *Independent* (13 January 2008) <http://www.independent.co.uk/arts-entertainment/art/reviews/fourth-plinth-national-gallery-london-769811.html> accessed 22 September 2009.

was never filled, even though it had been intended for an equestrian statue of William IV.[60] It remained vacant until 1999, when the first of three works commissioned by the Royal Society for the Encouragement of Arts, Manufactures and Commerce was installed. These installations were intended to generate discussion about what an appropriate and permanent modern artwork might be for the Plinth, bringing a sort of democracy to the choice of public art. However, the artworks generated such public interest that a new governing committee decided to use the space to display 'an ongoing series of temporary works of art commissioned from leading national and international artists.'[61]

When Livingstone became Mayor of London in 1999, he appointed a new Fourth Plinth Commissioning Group to select artworks for the site. For this Group, the Fourth Plinth program is foremost about using public space to engage with the arts; it is 'part of the vision for Trafalgar Square to be a vibrant, public space and to encourage debate about the place and value of public art in the built environment,'[62] rather than an explicit project of debate on national identity. In the process for commissioning an artwork, the Group selects a shortlist of artists who are asked to provide a maquette of their installation, from which a winner is chosen. In the selection process for 2012 and 2013, the Commissioning Group invited public comment prior to the announcement of the chosen installations. However, beyond this, there was no public reference to the process by which the winners were chosen, or the role that public opinion played in this process. The process is one in which 'elites' can be said to shape symbolism in the Square.

60   Mace, *Trafalgar Square*, 56.
61   City of Westminster, Planning applications sub-committee report on Trafalgar Square, London, WC2 (16 February 2012), 232 <http://transact.westminster.gov.uk/CSU/Planning%20Applications%20Committees/2008%20onwards /2012/07%20-%2016%20February/ITEM%2007-Trafalgar%20Square,%20WC2.PDF> accessed 15 March 2012.
62   Greater London Authority, 'The Fourth Plinth' <http://www.london.gov.uk/trafalgarsquare/around/4th_plinth.jsp> accessed 21 September 2010.

Even though the Commissioning Group focused on the value of the installations as contemporary artworks, for the public and the artists themselves, the sculptures have prompted a much wider discussion about British national identity, history and modernity in the context of Trafalgar Square. It would appear that that the site itself, its national symbolism, popular history and central urban location, has meant that the artworks cannot avoid commenting on British national identity. From 2005–2012, a series of artworks generated public discussion about the nation and its relationship with the individual, and I will discuss four of these before concluding this chapter.

The first statue installed under the new Fourth Plinth Commissioning Group was *Alison Lapper Pregnant* by artist Marc Quinn, which depicted the disabled Lapper eight months pregnant. Livingstone, who supported the scheme and endorsed the choice of works for display, described it as questioning 'our notions of who should be the subject of a statue or memorial'.[63] Quinn similarly compared it with the 'triumphant male statuary' in the Square and nearby Whitehall, and said that 'Nelson's Column is the epitome of a phallic male monument and I felt the square needed some femininity'.[64] The artistic strength of the statue was inseparable from its location in Trafalgar Square and its relationship to the statues on the other three plinths, all of whom personified martial power and imperial dominance.

A less positive interpretation was that the 'elites' who had chosen the statue had displayed shocking public contempt because 'public consensus' was neither sought nor valued; rather than allowing the vernacular or 'democratic' choice of heroes to emerge from below, the Commissioning Group was attempting to impose a top-down version of the nation through their

63   Fourth Plinth, 'Marc Quinn's Alison Lapper Pregnant – date announced for installation on Trafalgar Square's Fourth Plinth' Press release (15 June 2005) <http://www.london.gov.uk/fourthplinth/press/20050615.jsp> accessed 1 June 2010.

64   Nigel Reynolds, 'Whatever would Nelson think?', *The Telegraph* (16 September 2005) <http://www.telegraph.co.uk/news/uknews/4197485/Whatever-would-Nelson-think.html> accessed 3 June 2010.

defined use of the Square.[65] However, in constructing such a dichotomy, this argument overlooks the possibility that the Square might provide a site in which many different narratives can be accommodated. Nicolas Whybrow suggests that the statute was part of 'an evolving trend of positioning the public to view Trafalgar Square as a whole as a space of contesting narratives', the artwork acting as a 'resonant counterpoint to Nelson and company'.[66] *Alison Lapper Pregnant* can therefore be understood as both a challenge to the surrounding national narrative, and an opportunity to rewrite it.

This possibility was addressed in the summer of 2009 when sculptor Antony Gormley installed *One & Other*. Gormley's installation invited 2,400 people from all over Britain to stand on the Fourth Plinth for an hour for 100 consecutive days from 5 July 2009. Applicants were chosen at random by a lottery, but represented all the areas of Britain in the same proportion as their region's percentage of the national population. This meant, for example, that 207 Scots were asked to participate, and 333 people from the south-east of England.[67] The activities of the 'plinthers' were broadcast online via a webcam mounted on the plinth, therefore reaching a much wider audience than the one in Trafalgar Square itself.

For Gormley, this use of the plinth drew together many themes, including the power of the individual, his/her relationship with the national past and present, and the role of art in shaping public discussion about national identity. He described his project as an attempt to present a 'portrait of Britain made out of 2,400 hours of 2,400 people's lives'.[68] Furthermore, in creating this portrait, Gormley intended to contrast the participants with the figures on the other three plinths: 'We are celebrating the living,

65   Brendan O'Neill, 'Statue of limitations' *Spiked* (12 November 2007) <http://www.
      spiked-online.com/index.php?/site/article/4067/> accessed 7 June 2010.
66   Nicolas Whybrow, *Art and the City* (London: I.B. Tauris, 2011), 108.
67   T. Cornwell, 'Wanted: 207 Scots to stand on Trafalgar Square plinth', *The Scotsman*
      (27 February 2009) <http://news.scotsman.com/latestnews/wanted-207-Scots-to-
      stand.5022610.jp> accessed 7 July 2009.
68   Adrian Sooke, 'Antony Gormley's fourth plinth, Trafalgar Square', *The Telegraph* (26
      February 2009) <http://www.telegraph.co.uk/culture/4838343/Antony-Gormleys-
      Fourth-Plinth-Trafalgar-Square.html> accessed 7 July 2009.

and not the dead, the living who make up Britain in all its magnificence. We are creating a picture of Britain, and we don't yet know what the picture in composite will be.'[69] Just as Quinn had contextualised his work with the masculinity and 'heroism' in the other monuments in the Square, Gormley used these 'dead' historical figures to highlight the living individuals that collectively represented the present. Commentators picked up on this theme, with some describing *One & Other* as explicitly a project about British national identity, with the goal to use a narrative of the 'everyman/woman of Britain' to provide a form of popular democracy'.[70] The national portrait that the installation created was certainly diverse. Participants dressed variously as a town crier, a football referee, Britannia, a giant pigeon, a gorilla, human faeces, and a few wore nothing at all. Some of the participants explicitly referenced their immediate environment. Gerald Chong, for example, dressed as Godzilla and played tennis before destroying a cardboard model of London, including Nelson's Column and the Houses of Parliament.

As with *Alison Lapper Pregnant*, much of the media commentary in reaction to *One & Other* evaluated the project in terms of its place-specificity. In *The Telegraph*, Peter Whittle complained that it was 'spectacularly boring' and that it 'says nothing, illuminates nothing, and just adds to the mess which [...] is gradually eroding the grandeur of Britain's foremost public place'.[71] He complained that the installation was unsuitable for the Square, and diminished its national and historical significance. *The Guardian*'s Alex Needham, however, praised it as a 'life-affirming portrait revealing Britain's better side' that 'championed the little guy against the

69   Antony Gormley quoted in Charlotte Higgins, 'Fourth plinth gatecrashed, but Gormley is unfazed', *The Guardian* (6 July 2009) <http://www.guardian.co.uk/culture/charlottehigginsblog/ 2009/jul/06/ fourth-plinth-protest> accessed 7 July 2009.

70   Sooke, 'Antony Gormley's fourth plinth', 2009.

71   Peter Whittle, 'Antony Gormley's Fourth Plinth is a monumental bore', *The Telegraph* (12 July 2009) <http://blogs.telegraph.co.uk/news/ peterwhittle/100002979/ antony-gormleys-fourth-plinth-is-a-monumental-bore/> accessed 2 June 2010.

intimidating grandeur of the square's institutions'.[72] These accounts were place-specific, using the environment of the Square to frame both the installation and their commentary on it. Gormley also did this, discussing the contrast of the activities of the 'plinthers' with the built environment of the Square and its historic symbolism by describing how his work engaged with the history and national symbolism of the plinth. He expressed its impact in terms of a popular, democratic response to a dominant historical narrative:

> The square has its history as a place of national identity [...]. My project is about trying to democratise this space of privilege, idealisation and control. This is about putting one of us in the place of a political or military hero. It's an opportunity to use this old instrument of hierarchical reinforcement for something a little more [...] fun.[73]

Here Gormley casts his project as a reimagination of the nation *through* a reimagination of the site itself. The built and symbolic environment of Trafalgar Square was crucial for the project's success in commenting on modern identity, and was interwoven in both its historical representations and its contemporary everyday use. Furthermore, Gormley's characterisation of the existing environment of the Square as hegemonic and representative of 'official' narratives of identity serves to highlight his contrasting democratic and vernacular intervention. This degree of control necessary to complete the project – the computer-generated random selection of participants, the precise timing of the changeover of plinthers, and the requirement for participants to sign release forms – complicates Gormley's characterisation of his project as a challenge to the hegemonic power that he believed the Square represented and exercised. Miles made a similar point on the basis that the installation displayed a series of atomised individuals, each with their own message, rather than a coherent group narrative framed as 'national'.[74]

72   Alex Needham, 'Why the fourth plinth was a life-affirming portrait of Britain', *The Guardian* (9 October 2009) <http://www.guardian.co.uk/artanddesign/2009/oct/09/fourth-plinth-one-and-other-gormley?intcmp=239> accessed 8 June 2010.
73   Antony Gormley quoted in Sooke, *Antony Gormley's forth plinth*, 2009.
74   Miles, '*One & Other*', 358. For the performative aspects of Gormley's installation, see Whybrow, *Art and the City*.

As *One & Other* participants helped to build a 'composite picture' of Britain, their activities were shaped by the constraints of both the project's rules and the material environment of the Square. As with *Alison Lapper Pregnant*, the significance of this installation in terms of national identity – what tied it together into a larger (if loose) collective narrative – was the Square and its statues as a frame for meaning, a foil for comparison that underpinned the central animating theme: the contrast between the figure of the living individual on one plinth with the statues, representative of history, power and the nation, on the others.

The sculpture of Air Chief Marshal Sir Keith Park, installed just after *One & Other* also generated public discussion about national identity via a debate over the appropriate use of the Fourth Plinth. Sir Keith Park commanded the Royal Air Force in southeast England in 1940 during the Battle of Britain. A New Zealander by birth, he is credited as having held off the Luftwaffe, forcing Germany to reconsider its planned invasion of Britain, and his statue stood on the Plinth for six months from November 2009.[75] It was not commissioned specifically for the Fourth Plinth, unlike the other artworks, and its six-month tenure represented a compromise between the London government and campaigners who wanted a permanent statue of Park in the Square.

The contest over the permanency of the statue had begun the previous year. In March 2008, Terry Smith, the leader of the Sir Keith Park Memorial Campaign framed his support for the statue in terms of his views on the 'real' meaning of Trafalgar Square, based on a specific notion of national identity. Criticising the Fourth Plinth scheme, Smith said that 'the square was built to commemorate those who saved the nation and defended it. It was not intended as a contemporary art fair'.[76] During the 2008 mayoral

75   Greater London Authority, 'Sir Keith Park statue unveiled in Trafalgar Square' (4 November 2009) <http://www.london.gov.uk/media/ press_releases_mayoral/ sir-keith-park-statue-unveiled-trafalgar-square> accessed 9 June 2010.

76   Arifa Akbar, 'Boris and Benn form an unlikely plinth alliance', *The Independent* (8 March 2008) <http://www.independent.co.uk/arts-entertainment/art/news/boris-and-benn-form-an-unlikely-plinth-alliance-793146.html> accessed 2 September 2009.

election campaign, London's new mayor Boris Johnson had similarly framed the choice over what should fill the plinth as symbolic of larger national cultural and historical questions:

> I can go for a dead white male war hero, gloved, goggled, moustached, forged in traditional bronze and thereby – so I am warned – earn the odium of the entire liberal funkapolitan art world, or else I can continue to support the rotation of strange and wonderful works of contemporary art and enrage those who think these conversation pieces are out of keeping with Nelson's square and that a failure to install Sir Keith Park is a disservice to the memory of those who saved our country from tyranny in 1940 [...]. I say to the Keith Park campaigners 'some day your plinth will come'.[77]

In these examples, both Johnson and Smith describe the symbolism and purpose of Trafalgar Square in very narrow terms linked to wartime commemoration, characterising British national identity through the martial and masculine aspects of the past that are depicted in its statues and busts. Johnson traced the Square's saturation with a national historical narrative of martial victory, beginning with Nelson's naval victory, continuing through Havelock and Napier's imperial battles in India, and finally expressed in the busts of World War One naval commanders Beatty and Jellicoe. For Johnson, the question of what should be on the Fourth Plinth was based on the national history already represented in the Square. However, as the new London mayor, he also recognised that the Fourth Plinth's program of contemporary artworks was popular, and was reluctant to cancel it.

For other metropolitan officials, the choice of what occupied the plinth appeared to provide a possibility for a forward-looking discussion about the nature of national culture and identity, through a 'public debate about contemporary art'.[78] According to a Westminster Council planning report that rejected the possibility of a permanent memorial to Park on

---

77   Boris Johnson, 'The mayor's speech at the Royal academy annual dinner', (3 March 2008) <http://www.london.gov.uk/mayor/speeches/ 20080603-ra.jsp> accessed 2 September 2009.

78   Louise Jury, 'RAF hero snubbed as Boris keeps plinth art', (29 May 2008) <http:// www.thisislondon.co.uk/standard/article-23488227-raf-hero-snubbed-as-boris-keeps-plinth-art.do> accessed 9 February 2009.

the plinth, 'The fourth plinth is regarded as a site ideally suited for the display of provocative contemporary art'.[79] The same report found that a permanent statue of Park would be 'too representational and traditional'.[80] However, it was precisely this 'traditional' link to the history of World War Two that supporters of the statue invoked to buttress their claims. Sir Keith Park Memorial Committee chairperson Terry Smith, for example, used the history represented in the built environment of the Square to link Park's image with longstanding national narrative. When the statue was removed from the plinth in May 2010, Smith stated: 'Park's statue has fittingly sat beneath Nelson's Column – a memorial to another great commander who likewise defended Britain from invasion 135 years earlier'.[81] Writing in *The Independent*, Arifa Akbar expressed this as a cultural struggle over history that underlined Trafalgar Square's role as the symbolic heart of London, and by extension, the nation:

> Yesterday, more that 30 years after his death, Sir Keith Park was plunged into another tussle for the heart of the capital as a row over Trafalgar Square's empty fourth plinth sparked a furious standoff between some of the country's most eminent statesmen, historians and artists.[82]

Given that the statue was framed by both its supporters and detractors as symbolic of national identity, this is a reasonable interpretation. For supporters of the Fourth Plinth scheme, the Square was a valuable site for artworks that explored modern identity. For others, British national identity was best represented by a figure symbolising an instance of historical heroism, part of a longstanding narrative of British resistance to German

---

79 Murray Wardrop, 'Statue of Battle of Britain hero "not modern enough" for Trafalgar Square', *The Telegraph* (5 May 2009) <http://www.telegraph.co.uk/news/newstopics/howaboutthat/5279339/Statue-of-Battle-of-Britain-hero-not-modern-enough-for-Trafalgar-Square.html> accessed 10 June 2010.

80 *Ibid.*

81 'Sir Keith Park statue removed from Trafalgar Square', *BBC News* (5 May 2010) <http://news.bbc.co.uk/2/hi/uk_news/england/london/ 8662544.stm> accessed 9 June 2010.

82 Akbar, 'Boris and Benn', 2008.

attack during World War Two. This implicit reference to London's resilience during the Blitz was yet another instance of the impact that the Square's accreted symbolism can have. By forcing the national past and present into a contest with one another, supporters and detractors of the Park statue implied a contest amongst various narratives of national identity, suggesting that they were somehow mutually exclusive. However, other events and representations within the space had already shown that Trafalgar Square had the flexibility to use the past to frame the present, rather than necessarily forcing the two into conflict. Just as Massey has argued for the necessity of an 'unfinished and always becoming' understanding of place,[83] in reaching a compromise to allow Park's statue to stand for six months, London officials seemed to suggest that the Square was able to accommodate a range of national narratives.

After its allotted six months on the plinth, Park's statue was replaced by a sculpture by Yinka Shonibare MBE, a British-Nigerian artist with a record of works that have explored race, history and identity, particularly regarding the relationship between Britain and its former colonies. Shonibare's *Nelson's Ship in a Bottle* was installed on 24 May 2010. The sculpture was a 1:30 replica of Admiral Horatio Nelson's ship the HMS Victory, from which he commanded the Battle of Trafalgar in 1805. Its thirty-seven sails were set for battle and made of colourful and distinctive Dutch wax fabric, symbolic of the connections between Empire, naval power, global trade and British identity.[84]

One of the main themes in media reporting on Shonibare's artwork was how the sculpture reminded viewers that modern British multiculturalism has deep roots in a history of imperial expansion and conflict. For the artist, the connection between the past and the present was very important, and Shonibare linked this aspect of national identity specifically

---

83   Doreen Massey, *for space* (London: Sage, 2005), 59. Angharad Closs Stephens takes up the narrative of the Blitz in the context of the London bombings, but the argument that it foregrounds a white, English version of national identity also applies here.

84   Richard Dorment, 'Yinka Shonibare's Fourth Plinth, review', *The Telegraph* (24 May 2010) <http://www.telegraph.co.uk/culture/art/art-reviews/7760527/Yinka-Shonibares-Fourth-Plinth-review.html> accessed 2 June 2010.

to London: 'For me, it's a celebration of London's immense ethnic wealth, giving expression to and honouring the many cultures and ethnicities that are still breathing precious wind into the sails of the United Kingdom'.[85] *Ship in a Bottle* made direct reference to the Square's main monument, Nelson's Column, and at the unveiling of the sculpture, Shonibare was explicit about the site-specificity of his artwork: 'I think Nelson would be proud to see that his battle has had a significant effect on the lives of so many people. This piece celebrates the legacy of Nelson'.[86]

While *Ship in a Bottle* appeared to celebrate the preconditions for the modern British diversity that Shonibare values, it also hinted at darker aspects of imperialism, such as slavery and other forms of exploitation. Boris Johnson asked: 'Is it pro-empire? Is it anti-empire? This colourful and quirky take on our seafaring heritage provides a vivid contrast that intensifies the historic surroundings of Trafalgar Square'.[87] The sculpture seemed to highlight the historical setting, symbolically as well as visually. Charlotte Higgins also responded to the history underpinning the work, arguing that it recast a familiar London place, causing her to 'pay attention to the original reason for this square's existence'.[88] As with the other works discussed here, the relationship with its surrounding environment was central to the meaning of the installation, and it represented multiple and intertwining strands of history and identity, rather than mutually exclusive ones. The links between these historical narratives and modern London's multiculturalism were especially important to the artist. Shonibare recognised the flexible symbolic meaning of both the artwork and the his-

85   Yinka Shonibare quoted in Louise Jury, 'Fourth plinth art with a lot of bottle joins Nelson in Trafalgar Square', *The London Evening Standard* (24 May 2010) <http://www.thisislondon.co.uk/standard/article-23837422-fourth-plinth-art-with-a-lot-of-bottle-joins-nelson-in-trafalgar-square.do> accessed 1 June 2010.

86   'Nelson's ship in a bottle unveiled on Fourth Plinth', *The Telegraph* (24 May 2010) <http://www.telegraph.co.uk/culture/art/art-news/7758830/Nelsons-ship-in-a-bottle-unveiled-on-Fourth-Plinth.html> accessed 2 June 2010.

87   Boris Johnson quoted in Louise Jury, 'Fourth plinth'.

88   Charlotte Higgins, 'The fourth plinth: message in a bottle', *The Guardian* (24 May 2010) <http://www.guardian.co.uk/culture/charlottehigginsblog/2010/may/24/art-fourth-plinth> accessed 2 June 2010.

tory that inspired it, claiming that he was trying to be both celebratory and critical. His work suggested a narrative of a contemporary identity that was subject to the cultural and social tensions that form the legacy of Empire, but which also can produce a vibrant and unique metropolitan culture typified by tolerance. Overall, the Fourth Plinth, through its lack of a permanent statue or memorial, seems to ask for an important artwork to fill it, and repeatedly prompts a discussion of what is nationally important in contemporary Britain. In doing so it demonstrates the mutability and complexity of Britishness.

A subtle contest between different versions of the nation has played out within the confines of Trafalgar Square, evident in both the use of multi-culturalism to frame the events celebrating the Olympics and mourning the bombings in 2005, and in the official and media discussion of the Fourth Plinth scheme. While the pre-millennium structures in the Square have been treated as symbolic of an official, top-down narrative of the nation, the Fourth Plinth artworks were seen to augment, challenge or complicate this masculine and imperial story. In treating the contemporary installations as inextricable from their local environment, many commentators gestured towards a framing of national identity as discursive, multiple and complex. However, the built environment also limited this discourse, in some cases forcing the official into a contest with the vernacular. In terms of national identity, the significance of Shonibare's work was that it did not force the range of narratives expressed in the Square into a contest with each other, but instead presented them as inseparable, highly visible facets of modern British national identity. Recognising the ambivalence and multiple cultural and historical narratives within his work, Shonibare summed up his sculpture as 'a monument to live, and let live'.[89] Overall, the value of Trafalgar Square in illuminating some of the complexities of British national identity lies in the way it can draw together and make visible parallel but multiple narratives.

89   Yinka Shonibare quoted in Rachel Cooke, 'Yinka Shonibare: "I wanted to do a work connected to Trafalgar Square"', *The Guardian* (16 May 2010) <http://www.guard-ian.co.uk/artanddesign/ 2010/may/16/yinka-shonibare-fourth-plinth-trafalgar> accessed 2 June 2010.

CHAPTER 7

Conclusions

Tell them in England this: when first I stuck my head in the air, 'winched from a cockpit's tar and blood to my crow's nest over London, I looked down on a singular crowd moving with the confident swell of the sea. As it rose and fell every pulse in the estuary carried them quayward, carried them seaward.[1]

On Monday 8 April 2013, just as the manuscript of this book was being completed, former Prime Minister Margaret Thatcher died. Three-thousand people went to Trafalgar Square on the next Saturday to celebrate her death, drumming and chanting 'Maggie, Maggie, Maggie, dead, dead, dead'. They buried an effigy of the former leader made of recycled materials. 'Sparklers, party poppers and balloons' enlivened the gathering, which had been planned for years, and attracted people from across the country. Although there were a few arrests, 'the event was more like a party than a protest'.[2] As on many other occasions in the past, Trafalgar Square provided a means for people to make themselves heard on a national issue. One protester, a former miner from Newcastle, went to the gathering to oppose explicitly

---

1   From Jon Stallworthy, 'Epilogue to an Empire, 1600–1900, an ode for Trafalgar Day' in *Rounding the Horn: Collected Poems* (Manchester: Carcanet Press Ltd, 1998).

2   Tracy McVeigh and Mark Townsend, 'Thousands gather in Trafalgar Square to protest against Thatcher's legacy', *The Guardian* (13 April 2013) <http://www.guardian.co.uk/politics/2013/apr/13/margaret-thatcher-protest-trafalgar-square> accessed 19 April 2013; and Jessica Elgot, 'Margaret Thatcher Trafalgar Square Death Party Attended by Thousands', *Huffington Post* (13 April 2013) <http://www.huffington-post.co.uk/2013/04/13/margaret-thatcher-trafalgar-square-party_n_3077288.html> accessed 19 April 2013.

the notion that the country was in a state of mourning: 'We're absolutely furious at this image that is being presented on television, that the whole country is in mourning'. He had chosen the Square to make a visible statement about his version of history. Another participant thought it was 'important and cathartic for people to voice their dissent' against positive assessments of Thatcher's time as leader.[3] Suggesting limits to the protest, one attendee was 'skeptical of the political meaningfulness' but thought it 'positive that a lot of people have come together'.[4] They all appeared to regard the Square as an important site in which dissenting voices could be raised and, just as at other times in the Square's history, saw the site's value in expressing their views on a question of national significance.

In both her political life and her death, Trafalgar Square played a role in how Thatcher would be remembered. In 1990, the Square helped to precipitate her loss of the party leadership and the end of her political career through the Poll Tax Riot, a demonstration of violent discontent over her domestic policies and the social divisions they exacerbated. Given the role of the Square in her political history, a suggestion by a Conservative MP that she could be memorialised on the Fourth Plinth was almost ironic.[5] While Mayor Boris Johnson agreed that she deserved a statue in a prominent central London location, former mayor Ken Livingstone hinted that the Fourth Plinth was being reserved for a statue of the Queen, when the time came. The notion of putting Thatcher on the Plinth (or in any other prominent public location) was also a matter of concern on the basis that her statue could be subject to vandalism. This was not without good reason, as a previous statue of her on display in 2002 had been decapitated.[6]

3   McVeigh and Townsend, 'Thousands gather'.
4   *Ibid.*
5   Tom McTague, 'Margaret Thatcher statue plan for Trafalgar Square and bid to rename Falkland Islands' capital after her', *The Mirror* (10 April 2013) <http://www.mirror.co.uk/news/uk-news/margaret-thatcher-statue-plan-trafalgar-1823416> accessed 19 April 2013.
6   Matt Chorley, 'Boris: 'Put Thatcher statue next to Churchill', *The Daily Mail* (11 April 2013) <http://www.dailymail.co.uk/news/article-2307455/Margaret-Thatcher-

This book has argued that that the Square helps to construct Britishness in part through its value to minority organisations seeking both to raise publicity and to show themselves the scope of their own membership through physical occupation of the site. At the same time, such groups used the space of the Square to stake their own claim to the nation. At the time of writing, the Thatcher 'death party' was only the most recent example of this process. Minority groups, however, are not the only users of the Square: this book has discussed several examples of national events during which the Square was a central gathering place, its imperial symbolism subtly reinforcing official displays. For both the official and minority and vernacular expressions of identity within the Square, the site has acted as a proxy for the nation, allowing different users to construct different versions of Britishness, and stake their claim to the national narrative. Furthermore, historical uses of the Square have repeatedly been used to frame and legitimise later uses in a process of accreted symbolism that, according to Richard Williams, mirrors the built environment: 'The Square's history might therefore be said to be more the result of accident, accretion and history, in other words the usual ways of building in London'.[7]

It is important to remember, however, that the Square began in the nineteenth century as an unabashedly imperial space, at a time when London was 'a singular crowd moving with the confident swell of the sea'.[8] London was considered the heart of Empire, and Trafalgar Square was at the centre of the metropole, symbolically linked through a 'global sense of place' to places and people all over the Empire.[9] In this, the Square was part of a much larger landscape that stretched from the Docklands and City of London – with their direct financial and trading links to the Empire – to shops, clubs and palaces in the west, where decisions were made and goods were bought that had an impact on people almost unimaginably distant and different from those in central London. In the twenty-first century,

---

dead-Labour-says-Trafalgar-Square-fourth-plinth-statue-crass-triumphalism.html> accessed 19 April 2013.
7   Richard J. Williams, *The Anxious City* (London: Routledge, 2004), 142.
8   Stallworthy, 'Epilogue to an Empire'.
9   Massey, 'A Global sense of place'.

Nelson's Ship in a Bottle provided visual continuity between imperial and post-colonial Britain, with an installation that spoke directly to questions of modern British multiculturalism, the uses of history to interpret and produce national identity in the present, and Trafalgar Square as a site of accreted symbolism that has been used repeatedly to buttress official national narratives.[10] The artist Shonibare himself described the artwork as an exploration of the diverse cultural influences implied by British national identity, rather than identifying it with a single totalising narrative: 'I love what you could call "vindaloo Britishness". It's a mixed-up thing. You hear it in British music, and you taste it in British food. This purity notion is nonsense, and I cherish that'.[11] With Shonibare's work, the Square's many possible narratives were shown to have sprung from a complex past with global reach reflected in the site's built environment.

The built environment surrounding the Square has also contributed to its significance and meaning. Royal ceremonial events passed the Square going to or from nearby Buckingham Palace, St Paul's Cathedral and Westminster Cathedral, and members of the crowd enjoyed the experience and spectacle of being part of the throng on such occasions, even when they were unable to catch a glimpse of the passing dignitaries and 'there was nothing to see but crowds'. Protesters gathering in the Square were very close to Parliament and Downing Street, and to larger rallies in Hyde Park. From almost the moment it was built, the Square has been a place for the disempowered or disenfranchised to communicate directly with government; to organise themselves and show their numbers; to air their grievances; and to demand change by making themselves heard, both through the media and through physical occupation of the site. Protest has always been an important part of the Square's history, and remains central to how it is popularly understood.

The record of political change directly attributable to protest in the Square, however, is not unambiguous. For example, although the Suffragettes used the site repeatedly before the First World War, it was not until

10   See Dwyer, 'Symbolic accretion'.
11   In Cooke, 'Yinka Shonibare'.

1918 that limited female enfranchisement was achieved, by which time direct protest had long ceased. On the other hand, the Poll Tax Riot in 1990 represented the culmination of national resistance to an unpopular policy that ended Thatcher's Prime Ministership. Therefore, instead of thinking of the Square as a site through which 'the people' can demand and attain change, it is more useful to recognise how minority or less-powerful groups used it to stake their claim in the national story, and to argue for reimagined versions of Britishness (alongside other identities) in which their interests were visible. The Square can thus be understood as a proxy for the nation, a debating chamber in which many different national narratives can be contested. John Parkinson makes a similar argument, linking democracy itself to the use of public space for narrating political argument, making public claims and expressing public displeasure.[12]

In this sense, control of use or access to the Square is a highly political question and allowing only certain groups access to it narrows the range of claims that can be made on Britishness. Authorities have implicitly recognised this in the groups they have prevented from using the site. In 1972, for example, protests related to Northern Ireland were specifically prohibited. The Department of the Environment, which had administrative responsibility for the Square, said that a recent IRA bombing at Aldershot and the 'escalation in public feelings over Northern Ireland' meant that there was dangerous potential for violence.[13] Meetings were also limited at other times, such as during the First World War under the Defence of the Realm Act. Allowing access was also powerful, such as the permission to use the Square that was given to Oswald Mosley and his supporters in 1962 on the basis of 'free speech' (the meeting was violently disrupted). More recently, in amended bye-laws that came into force in March 2012, the use of tents and sleeping bags in the Square require written permission from the Mayor of London, as does using amplification equipment. Presumably this is a response to particular tactics, such as those used by people objecting to rises in university tuition fees and education cuts in

12   John Parkinson, *Democracy and Public Space* (Oxford: OUP, 2012), 204–205.
13   Mary Holland, 'Trafalgar Square rally Ban', *The Observer* (19 March 1972), 3.

November 2011; this group briefly set up camp in Trafalgar Square in a protest that replicated the tactics of the Occupy movement which had set up a similar camp a short distance away at St Paul's Cathedral.[14]

In such instances, the Square is a protest site; it can also be used for more gentle purposes. For example, it showed a much quieter side in August 2008, with banners announcing the annual Trafalgar Square Festival, promising 'amazing outdoor performances and live coverage from the Beijing Olympic Games'. These Games were indeed shown on a big screen mounted underneath Nelson's Column that displayed Team GB working its way towards fourth place in the medal tally. When I visited, many people were seated on the steps facing the screen, turning the space into a temporary, impromptu amphitheatre, but they were watching the big screen casually, still eating, talking and moving around the Square. Children broke away from their parents to run across the pavement and chase the few hopeful pigeons. As people moved across the open spaces of the Square, they stopped to eat, take photographs, or absorb their surroundings, diverted briefly to take in the atmosphere. The personal and intimate activities of eating or child-minding coincided with the official national narrative represented by the broadcast of Olympics coverage. There was a sense of possibility, of spontaneity, of temporary, pedestrian diversion, as people paused in public space. This is another important aspect of the Square that helps to reinforce its role as a site of imagining the nation. On most days, it is an everyday space, a backdrop for normal life, not one of spectacle, violence or explicit national demonstration.

Richard Williams argues that the 'civility' that was the goal of the Square's redesign in 2003 has obscured its history as a site of conflict and protest, creating a faux-Italian 'piazza' that does not suit the London climate or culture.[15] He suggests that London has been left with a bloodless heart, pleasant enough for tourists or middle-class visitors to the National Gallery,

---

14   Richard Alleyne, Mark Hughes, Victoria Ward and James Orr, 'London protests: police put a stop to Trafalgar Square "tent city"', *The Telegraph* (9 November 2011) <http://www.telegraph.co.uk/education/universityeducation/8880180/London protests-police-put-a-stop-to-Trafalgar-Square-tent-city.html> accessed 24 July 2012.

15   See Williams, *The Anxious City*, Chapter 6.

but irrelevant and potentially exclusive of the majority of Londoners and Britons. Certainly, the control of the space has become much tighter throughout its history. While there have almost always been some controls or regulations over what happens there, the quotidian uses have become more limited to tourism or transit, with the space constantly patrolled by Heritage Wardens. Its redesign also saw it refashioned as an official site of multiculturalism. In an argument for the importance of London's multiculturalism to its contemporary identity, White points out that:

> Within twenty years of the first mass arrivals most Londoners accepted that their city had changed forever, and for good. With another twenty, London was lauded as one of the most tolerant and successful of the world's multicultural societies. It had proved in past a melting pot [...] but even more a great constellation of the world's people retaining their own identity in a single city.[16]

For the GLA, the Square appears to have a very clearly defined role as a 'multicultural' space, and this policy points to an overall official emphasis on urban cohesion that had come to dominate the official narrative of Trafalgar Square by the time the Fourth Plinth scheme began. This demonstrates official attempts to shape the narratives that the Square represents, even as its statuary and memorials remain more or less the same. This policy has seen the Square used for festivals specific to London's diverse cultural communities, including Eid, the festival marking the end of the Muslim fasting month of Ramadan, and Diwali, the Hindu festival of light. On the Saturday in October 2008 when 'Eid in the Square' took place, for example, Trafalgar Square was crowded with families, but was no longer predominantly a tourist destination. It had become a more enclosed space, more demarcated from its surroundings than usual. It contained the same visual elements, but these had been transformed into background elements for a modern celebration of urban diversity. The UK faith-based representative group the Muslim Council of Britain emphasised the role of the event as an opportunity to create individual connections between Muslim and non-Muslim co-nationals, both within the Square and nationally, in

---

16 White, *London in the 20th Century*, 405.

a 'landmark event that will benefit not only all Londoners, but the rest of the nation [...]. We hope this will be replicated across the country, where Muslims will invite fellow citizens in their celebration'.[17] In his speech opening the celebration, London Mayor Boris Johnson highlighted an individual relationship to Islam: 'exactly 100 years ago, in 1908, my father's father's father came to south London. He was a Muslim and he knew the Koran off by heart in Arabic [...]. He would be absolutely amazed to discover that his great-grandson had become mayor of London'.[18] In invoking his great-grandfather, Johnson appeared to link the event in the Square to a history of British multiculturalism. Furthermore, his legitimisation of the use of the Square for a celebration of Eid helped to recognise Muslims as British. This example illustrates how the Square can have flexible meanings despite its established topography. Throughout the site's history, a wide range of groups has claimed national belonging by including themselves in who the Square is 'for'. Officially recognised and sanctioned, these events contribute to the flexibility of the Square's meanings and help to build a place-based national narrative that reframes the past to assert the belonging of diverse groups in the national British present.

Overall, this book has been an argument for the powerful role that national places can play in the construction, maintenance or transformation of narratives of national identity. It has traced detailed examples of how different groups have used Trafalgar Square to imagine themselves as part of the British national community, and how the prominence of the Square has helped to legitimise users' claims. In doing so, the book seeks to include national public places in the roll-call of 'institutions' that help to create and strengthen national narratives, a list that incudes Anderson's maps, museums and censuses and Billig's postage stamps, weather broadcasts and limp flags.[19] As Dwyer and Parkinson both point out, monumental

17   Muslim Council of Britain, 'Muslims Celebrate Eid in the Capital', (11 October 2008) <http://www.mcb.org.uk/media/ presstext.php?ann_id=314> accessed 4 August 2010.

18   'Trafalgar Square hosts Eid event', *BBC News* (11 October 2008) <http://news.bbc. co.uk/ 2/hi/uk_news/england/london/7665154.stm> accessed 8 September 2009.

19   See Anderson, *Imagined Communities* and Billig, *Banal Nationalism*.

places are 'political resources' with important implications for democratic practice, and the groups that can exploit this resource are wide-ranging.[20] Public place is flexible and able to be occupied by many different groups, whose activities shift or transform its meanings for subsequent users.

As such, the history of Trafalgar Square supports characterisations of national identity as flexible and discursive, and able to accommodate multiple narratives. In doing so, it demonstrates the interaction between the official, or 'top-down', and the vernacular, or 'from-below', that Hobsbawm and Ranger remind us is important in defining the limits of the national community.[21] The relationship between the state and the people is made tangible in public space, including the state's capacity to manage or even incorporate contest, dissent and protest. However, Massey and Nora both remind us that if 'place' is a spatial category, it is also a temporal one.[22] The role of national *lieux de mémoire* in presenting a version of the past for use in the present provides a brake on the discursive process of national identity formation. Place history – both in terms of what place represents and what has happened in a given place – helps to shape how groups use a site and what rhetorical advantage they can gain from it. The accretion of symbolism over time and the way place can 'make the past come to life in the present and thus contribute to the production and reproduction of social meaning' are both central to its role in helping to construct national identity.[23] This book has sought to provide a detailed case study of how these intertwined processes work, and how national identity has been imagined and narrated in one of the best-known public places in the world, Trafalgar Square.

20 Dwyer, 'Accreted symbolism' and Parkinson, *Democracy and Public Space*.
21 Hobsbawm and Ranger, eds, *The Invention of Tradition*.
22 Massey, 'Places and Their Pasts' and Nora, 'Les *Lieux de Mémoire*'.
23 Dwyer, 'Accreted symbolism' and Cresswell, *Place*.

Bibliography

Archival sources

Anti-Apartheid Movement Archive, Bodleian Library of Commonwealth and African Studies, University of Oxford.

London School of Economics Archives, Campaign for Nuclear Disarmament Collection.

National Archives: CAB/65/50/16, Arrangements for Celebrating the End of Hostilities in Europe, 27 April 1945.

National Archives: CAB/66/65/19, Celebration of VE-Day, 25 April 1945.

National Archives: CAB/65/50/22, Arrangements in Connection with the End of Hostilities in Europe, 7 May 1945.

National Archives: CAB/24/58, War Cabinet minutes, 17 July 1918.

National Archives: CAB/23/8, War Cabinet minutes, 14 November 1918.

National Archives CAB/24/70: War cabinet memorandum from Commissioner of Police Macready, 18 November 1918.

National Archives: HO 186/2050, Lighting restrictions, Arrangements for celebrating the cessation of hostilities with Germany, Flood Lighting, 1945. Second Report of the Interdepartmental Conference.

National Archives: MEPO 2/1556, Trafalgar Square Meetings: 'Free Speech Defence Committee', 1913.

National Archives: MEPO 2/7218, Control of meetings in Trafalgar Square: prohibition of any assembly or procession in the vicinity of the Houses of Parliament: Home Office and police instructions, 1918–1948.

Mass Observation Archive FR 2263 1945. With thanks to the Trustees of the Mass Observation Archive, University of Sussex.

House of Commons Hansards

House of Lords Hansards

London 2012 – www.london2012.com

Greater London Authority – www.london.gov.uk

British Pathé – www.britishpathe.com

Newspapers and news websites

The following sources are UK titles unless otherwise indicated or evident.

The Advocate [Burnie, Tasmania]
BBC News Online
Daily Express
Daily Mail
Daily Mirror
Daily News
Daily Telegraph and Morning Post
Evening Standard
The Guardian
Huffington Post
The Independent
Illustrated London News
LA Times [Los Angeles]
London Evening Standard
Financial Times
The Manchester Guardian
Mirror
Observer
The Scotsman
The Suffragette
Sun
Telegraph
The Times

Other primary sources

The Australians' Guide Book to London, High Commissioner, Australia House, 1924.
The Australians' Guide Book to London, High Commissioner, Australia House, 1930.
Baedeker, Karl, *London and its environs, including excursions to Brighton, the Isle of Wight, etc* (London: Dulau and Co, 1883).

Blair, Tony, 'PM's comments on 2012 Olympic Games', 6 July 2005 <http://www.number10.gov.uk/Page7832> accessed 13 September 2010.

Brown, Gordon, 'The Future of Britishness', Speech to the Fabian Future of Britishness conference, 14 January 2006 <http://www.fabian-society.org.uk/press_office/news_latest_all.asp? pressid=520> accessed 21 May 2007.

Carlton, Mary, 'Memories of VE Day', WW2 People's War, http://www.bbc.co.uk/history/ww2peopleswar/stories/29/a4143629.shtml> accessed 10 April 2013.

City of Westminster, Planning applications sub-committee report on Trafalgar Square, London, WC2 (16 February 2012) <http://transact.westminster.gov.uk/CSU/Planning%20Applications%20Committees/2008%20onwards /2012/07%20-%2016%20February/ITEM%2007-Trafalgar%20Square,%20WC2.PDF> accessed 15 March 2012.

The Coronation of their Majesties King George VI and Queen Elizabeth: Official Souvenir Programme, King George's Jubilee Trust, 1937.

From the Four Corners, Denham and Pinewood Studios, 1941.

Livingstone, Ken, 'London United', 14 July 2005 <http://www.youtube.com/watch?v=6BSIBPsbL9c> accessed 4 March 2009.

'Loudspeakers in Trafalgar Square, London, England, UK, 1941', Imperial War Museum, < http://www.iwm.org.uk/collections/item/object/205195856> accessed 10 April 2013.

Murray, E.G., *Personal diary, 1895–1918* (Murray, Utah: Family Heritage Publishers, 2007) [the Women's Library, London Metropolitan University].

Muslim Council of Britain, 'Muslims Celebrate Eid in the Capital', 11 October 2008 <http://www.mcb.org.uk/media/ presstext.php?ann_id=314> accessed 4 August 2010.

National Statistics Online, 'London: population and migration', 8 June 2010 <http://www.statistics.gov.uk/cci/nugget.asp?id=2235> accessed 23 August 2010.

O'Neill, Brendan, 'Statue of limitations', 12 November 2007 <http://www.spiked-online.com/index.php?/site/article/4067/> accessed 7 June 2010.

Palliser, Edith, 'Card dated 19th June 1908 to Miss P. Strachey from Edith Palliser from Amsterdam congratulating her on the success of the Procession of June 13, 1908'. Autograph Letters Collection regarding the 'suffragist procession', 13 June 1908. Microfiche Box 1, Vol 1 (A–J), June 1908, 9/01/0407 (ALC/407), the Women's Library, London Metropolitan University.

Royal British Legion, 'Memories of VE Day' <http://www.britishlegion.org.uk/remembrance/ve65/memories-of-ve-day> accessed 27 April 2010.

Secondary sources

Alibhai-Brown, Yasmin, 'Muddled leaders and the Future of British National Identity', *The Political Quarterly* 71/1 (2000), 26–30.

Alibhai-Brown, Yasmin, *Imagining the New Britain* (New York: Routledge, 2001).

Allen, John, 'Ambient power: Berlin's Potsdamer Platz and the seductive logic of public spaces', *Urban Studies* 43/2 (2006), 441–455.

Anderson, Benedict, *Imagined Communities* (London: Verso, 1991 [1983]).

Anderson, Ben and Wylie, John, 'On Geography and Materiality', *Environment and Planning A* 41 (2009), 318–335.

Anonymous, *Poll Tax Riot: 10 hours that shook Trafalgar Square* (London: Acab Press, 1990).

Atkinson, Diane, *Suffragettes in Purple, White and Green: London 1906–1914* (London Museum of London, 1992).

Bell, Amy Helen, 'Landscapes of Fear: Wartime London, 1939–1945', *Journal of British Studies* 48 (2009), 153–175.

Best, Geoffrey, *Churchill: A Study in Greatness* (London: Hambeldon and London, 2001).

Bhabha, Homi, 'Introduction', *Nation and Narration* (London: Routledge, 1990).

Billig, Michael, *Banal Nationalism* (London: Sage, 1995).

Bloom, Clive, *Violent London* (Basingstoke: Palgrave Macmillan, 2010).

Briggs, R.C.H., 'Morris and Trafalgar Square', *Journal of William Morris Studies* (Winter 1961), 28–31, <http://www.morrissociety.org/publications/JWMS/W61.RCHB.pdf> accessed 25 March 2013.

Burkett, Jodi, 'Re-defining British morality: "Britishness" and the Campaign for Nuclear Disarmament 1958–68', *Twentieth Century British History* 21/2 (2010), 184–205.

Butler, David, Adonis, Andrew and Travers, Tony, *Failure in British Government: The Politics of the Poll Tax* (Oxford: Oxford University Press, 1994).

Cabell, Craig and Richards, Allan, *VE Day: A Day to Remember* (Long Preston: Magna Large Print Books, 2005).

Calder, Angus, *The People's War: Britain 1939–1945* (London: Pimlico, 2008[1969]).

Calhoun, Craig, *Nationalism* (Buckingham: Open University Press, 1997).

Campbell, Alexander, *It's Your Empire* (London: Left Book Club, 1945).

Cannadine, David, 'The Context, Performance and Meaning of Ritual: The British Monarchy and the "Invention of Tradition", c. 1820–1977' in Hobsbawm, Eric and Ranger, Terence, eds, *The Invention of Tradition* (Cambridge: Cambridge University Press, 1983).

Chakravarty, Gautam, *The Indian Mutiny and the British Imagination* (Cambridge: CUP, 2006).

Cherry, Deborah, 'Statues in the Square: Hauntings at the Heart of Empire', *Art History* 29/4 (2006), 660–697.

Closs Stephens, Angharad, '"Seven Million Londoners, One London": National and Urban Ideas of Community in the Aftermath of the 7 July 2005 Bombings in London', *Alternatives* 32 (2007), 155–176.

Cohen, Anthony, 'personal nationalism – a Scottish view of some rites, rights, and wrongs' *american ethnologist* 23/4 (1996), 802–815.

Colley, Linda, *Britons: Forging the Nation, 1707–1837* (London: Pimlico, 1992).

Conekin, Becky, *'The autobiography of a nation': The 1951 Festival of Britain* (Manchester: Manchester University Press, 2003).

Connelly, Mark, *We Can Take It! Britain and the Memory of the Second World War* (London: Longman, 2004).

Corthorn, Paul and Davis, Jonathan, *The British Labour Party and the Wider World: Domestic Politics, Internationalism and Foreign Policy* (London: Tauris, 2008).

Cresswell, Tim, *Place: A short introduction* (Oxford: Blackwell, 2004).

Day, Barry, ed., *The Essential Noël Coward Compendium: The Very Best of His Work, Life and Times* (London: Methuen Drama, 2009).

Day, Graham and Thompson, Andrew, *Theorizing Nationalism* (Basingstoke: Palgrave Macmillan, 2004).

Deacon, David and Golding, Peter, *Taxation and Representation: The Media, Political Communication and the Poll Tax* (London: John Libbey, 1994).

DeLyser, Dydia, '"Thus I salute the Kentucky Daisey's claim": gender, social memory, and the mythic West at a proposed Oklahoma monument', *Cultural Geographies* 15 (2008), 63–94.

Dennis, Richard, *Cities in Modernity: Representations and Productions of Metropolitan Space, 1840–1930* (Cambridge: Cambridge University Press, 2008).

Driver, Felix and Gilbert, David, 'Heart of Empire? Landscape, space and performance in imperial London', *Environment and Planning D: Society and Space* (16) 1998, 11–28.

Dwyer, Owen, 'Symbolic accretion and commemoration' *Social and Cultural Geography* 5/3 (2004), 419–425.

Eade, Charles, ed., *Victory: War Speeches by the Right Hon. Winston S. Churchill, O.M., C.H., M.P.* (Melbourne: Cassell and Company Ltd, 1945).

Edensor, Tim, *National Identity, Popular Culture and Everyday Life* (Oxford: Berg, 2002).

Fieldhouse, Roger, *Anti-apartheid: a history of the movement in Britain: a study in pressure group politics* (London: Merlin, 2005).

Fletcher, Ian Christopher, Mayhall, Laura E. Nym, Levine, Philippa, eds, *Women's Suffrage in the British Empire: Citizenship, Nation and Race* (Abingdon: Routledge, 2000).

Forest, Benjamin, Johnson, Juliet and Till, Karen, 'Post-totalitarian national identity: public memory in Germany and Russia', *Social and Cultural Geography* 5/3 (2004), 357–380.

Fryer, Peter, *Staying Power: The History of Black People in Britain* (London: Pluto-Press, 2010 [1984]).

Gardiner, Juliet, *The Thirties: An Intimate History* (London: HarperPress, 2010).

Garnett, Mark, *From Anger to Apathy: The Story of Politics, Society and Popular Culture in Britain since 1975* (London: Vintage, 2007).

German, Lindsey and Rees, John, *A People's History of London* (London: Verso, 2012).

Gilbert, David '"London in all its glory – or how to enjoy London": guidebook representations of imperial London', *Journal of Historical Geography* 25/3 (1999), 279–297.

Gillis, John R., *Commemorations: The Politics of National Identity* (Princeton: Princeton University Press, 1994).

Gilroy, Paul, *After Empire: Melancholia or Convivial Culture?* (Abingdon: Routledge, 2004).

Gregory, Adrian, *The Last Great War: British Society and the First World War* (Cambridge: Cambridge University Press, 2008).

Guibernau, Montserrat, *The Identity of Nations* (Cambridge: Polity Press, 2007).

Gurney, Christabel, '"A Great Cause": The Origins of the Anti-Apartheid Movement, June 1959–March 1960', *Journal of Southern African Studies* 26/10 (2000), 123–144.

Hagen, Joshua and Ostergren, Robert, 'Spectacle, architecture and place at the Nuremberg party rallies: projecting a Nazi vision of past, present and future' *Cultural Geographies* 13 (2006), 157–181.

Halbwachs, Maurice (trans. by Lewis A. Coser) *On Collective Memory* (Chicago: University of Chicago Press, 1992).

Hargreaves, Roger, *Trafalgar Square Through the Camera* (London: National Portrait Gallery, 2005).

Harris, Jose, *Private Lives, Public Spirit: Britain 1870–1914* (London: Penguin Books, 1994).

Harvey, David, 'Contested Cities: Social Process and Spatial Form' (1997) in LeGates, Richard and Stout, Frederic, eds, *The City Reader* (Routledge: New York, 2003).

Heffernan, Michael, 'For ever England: the Western Front and the politics of remembrance in Britain' *Cultural Geographies* 2 (1995), 293–323.

Hobsbawm, Eric and Ranger, Terence, eds, *The Invention of Tradition* (Cambridge: Cambridge University Press, 1983).

Hobsbawm, Eric, *The Age of Empire: 1875–1914* (London: Abacus, 1987).

Hoelscher, Steven and Alderman, Derek, 'Memory and place: geographies of a critical relationship' *Social and Cultural Geography* 5/3 (2004), 347–355.

Hood, Jean, *Trafalgar Square: A Visual History of London's Landmark Through Time* (London: Batsford, 2005).

Hutchinson, John, *Nations as Zones of Conflict* (London: Sage, 2005).

Hutchinson, John and Smith, Anthony, eds, *Nationalism* (Oxford: OUP, 1994).

Iveson, Kurt, *Publics and the City* (Oxford: Blackwell, 2007).

Jennings, Humphrey and Madge, Charles, eds, *Mass-Observation Day Survey: May 12 1937* (London: Faber and Faber, 1937).

John, Angela and Eustance, Claire, eds, *The Men's Share? Masculinities, Male Support and Women's Suffrage in Britain, 1890–1920* (London and New York: Routledge, 1997).

Johnson, Nuala, 'Sculpting heroic histories: celebrating the centenary of the 1798 rebellion in Ireland', *Transactions of the Institute of British Geographers* 19/1 (1994), 78–93.

Johnson, Nuala, 'Cast in stone: monuments, geography and nationalism', *Environment and Planning D: Society and Space* 13 (1995), 51–65.

Johnson, Nuala, 'The contours of memory in post-conflict societies: enacting public remembrance of the bomb in Omagh, Northern Ireland', *Cultural Geographies* (2011), published online 10 November.

Jones, Rhys, 'Relocating nationalism: on the geographies of reproducing nations', *Transactions of the Institute of British Geographers* 33 (2008), 319–334.

Jones, Rhys and Merriman, Peter, 'Hot, banal and everyday nationalism: bilingual road signs in Wales' *Political Geography* 28/3(2009), 164–173.

Jones, Rhys and Fowler, Carwyn, 'Placing and Scaling the nation' *Environment and Planning D: Society and Space* 25 (2007), 332–354.

Jorgensen-Earp, Cheryl, ed., *Speeches and Trials of the Militant Suffragettes: The Women's Social and Political Union, 1903–1918* (London: Associated University Presses, 1999).

Kelly, Katherine, 'Seeing Through Spectacles: The Woman Suffrage Movement and London Newspapers, 1906–1913', *European Journal of Women's Studies* 11/3 (2004), 327–353.

Kincaid, Andrew, *Postcolonial Dublin: Imperial Legacies and the Built Environment* (Minneapolis: University of Minnesota Press, 2006).

King George's Jubilee Trust, *The Coronation of their Majesties King George VI and Queen Elizabeth: Official Souvenir Programme* (London: Odhams Press Ltd, 1937).

Kynaston, David, *Austerity Britain 1945–51* (London: Bloomsbury, 2007).

Lawrence, Jon, *Electing Our Masters: The Hustings in British Politics from Hogarth to Blair* (Oxford: OUP, 2009).

Lees, Loretta, 'Rematerialising geography: the "new" urban geography', *Progress in Human Geography* 26/1 (2002), 101–112.

Lewis, David Stephen, *Illusions of Grandeur: Mosley, Fascism and British Society, 1931–1981* (Manchester: Manchester University Press, 1987).

Longmate, Norman, *When We Won the War: The Story of Victory in Europe, 1945* (London: Hutchinson & Co, 1977).

Lowenthal, David, *The Past is a Foreign Country* (Cambridge: Cambridge University Press, 1985).

Mace, Rodney, *Trafalgar Square: Emblem of Empire* (London: Lawrence and Wishart, 2005 [1976]).

MacKenzie, John, ed., *Imperialism and Popular Culture* (Manchester: Manchester University Press, 1986).

Massey, Doreen, 'A Global Sense of place' *Marxism Today* (June 1991), 23–27.

Massey, Doreen, 'Politics and Space/Time' *New Left Review* 196 (Nov-Dec 1992), 65–84.

Massey, Doreen, 'Places and their Pasts', *History Workshop Journal* 39 (1995), 182–192.

Massey, Doreen, *for space* (London: Sage, 2005).

Massey, Doreen, *World City* (London: Polity, 2007).

Merriman, Peter; Revill, George; Lorimer, Hayden; Matless, David; Rose, Gillian; Wylie, John; Cresswell, Tim, 'Landscape, mobility, practice' *Social and Cultural Geography* 9/2 (2008), 191–212.

Miles, Malcolm, 'One & Other: a picture of the nation in a period of cosmopolitanism?' *The Journal of Architecture* 16/3 (2011), 347–363.

Miller, David, *On Nationality* (Oxford: OUP, 1995).

Miller, Russell, *Ten Days in May: The People's Story of VE Day* (London: Michael Joseph, 2007 [1995]).

Modood, Tariq, *Multiculturalism: A Civic Idea* (Cambridge; Polity, 2013).

Nash, Catherine 'Performativity in practice: some recent work in cultural geography' *Progress in Human Geography* 24 (2000), 653–664.

Nora, Pierre, 'Between Memory and History: Les Lieux de Mémoire' *Representations* 26 (Spring 1989), 7–24.

Olick, Jeffrey, *States of Memory* (Durham and London: Duke University Press, 2003).

Orwell, George, 'England Your England' in *Inside the Whale and Other Essays* (London: Penguin Books, 1957).

Özkırımlı, Umut, *Contemporary Debates on Nationalism: A Critical Engagement* (Basingstoke: Palgrave Macmillan, 2005).

Pankhurst, Christabel, *Unshackled: The Story of How We Won the Vote* (London: Hutchinson, 1959).

Parkinson, John R., *Democracy and Public Space: The Physical Sites of Democratic Performance* (Oxford: OUP, 2012).

Percy, W.S., *The Empire Comes Home* (London: Collins, 1937).

Perry, Matt, *The Jarrow Crusade: Protest and Legend* (Sunderland: University of Sunderland Press, 2005).

Phillips, Melanie, *The Ascent of Woman: A History of the Suffragette Movement and the Ideas Behind It* (London: Abacus, 2003).

Porter, Bernard, *The Absent-Minded Imperialists: Empire, Society and Culture in Britain* (Oxford: OUP, 2004).

Porter, Roy, *London: A Social History* (Cambridge, Mass: Harvard University Press, 2001).

Powell, David, *The Edwardian Crisis: Britain 1901–1914* (London: Macmillan Press Ltd, 1996).

Purvis, June and Stanley Holton, Sandra, *Votes For Women* (New York: Routledge, 2000).

Purvis, June, *Emmeline Pankhurst: A Biography* (London: Routledge, 2002).

Rappaport, Erika, *Shopping for Pleasure: Women in the Making of London's West End* (Princeton, New Jersey: Princeton University Press, 2000).

Renan, Ernst, 'Qu'est-ce qu'une nation?' in Hutchinson, John and Smith, Anthony, eds, *Nationalism* (Oxford: Oxford University Press, 1994).

Robb, George, *British Culture and the First World War* (Basingstoke: Palgrave, 2002).

Rose, Mitch, 'Landscape and labyrinths', *Geoforum* 33 (2002), 455–467.

Rose, Sonya, *Which People's War? National Identity and Citizenship in Britain 1939–1945* (Oxford: OUP, 2003).

Said, Edward, 'Invention, Memory, and Place', *Critical Inquiry* 26/2(2000), 175–192.

Sandbrook, Dominic, *State of Emergency: The Way We Were; Britain, 1970–1974* (London: Penguin, 2011).

Schama, Simon, *A History of Britain: The Fate of Empire 1776–2000* (London: BBC Worldwide Ltd, 2002).

Schneer, Jonathan, *London 1900: The Imperial Metropolis* (New Haven: Yale University Press, 1999).

Sheppard, Francis, *London: A History* (Oxford: OUP, 1998).

Simmonds, Alan, *Britain and World War One* (London: Routledge, 2012).

Smith, Anthony, 'The limits of everyday nationhood', *Ethnicities* 8 (2008), 563–573.

Staiger, Uta, 'Cities, citizenship, contested cultures: Berlin's Palace of the Republic and the politics of the public sphere', *Cultural Geographies* 16 (2009), 309–327.

Stallworthy, Jon, *Rounding the Horn: Collected Poems* (Manchester: Carcanet Press Ltd, 1998).

Stedman Jones, Gareth, *Outcast London: A Study in the Relationship Between Classes in Victorian Society* (Oxford: Clarendon Press, 1971).

Stedman Jones, Gareth, 'The "cockney" and the nation, 1780–1988' in Feldman, David and Stedman Jones, G., eds, *Metropolis London: Histories and Representations since 1800* (London: Routledge, 1989).

Susman, Tony, 'Sharpeville Massacre' in 'Letters to the Editor', *The Times* (2 April 1970).

Szreter, Simon, *Fertility, Class and Gender in Britain, 1860–1940* (Cambridge: Cambridge University Press, 1996).

Thompson, Andrew, 'Nations, national identities and human agency: putting people back into nations', *The Sociological Review* 49/1 (2001), 18–32.

Thompson, Andrew, *The Empire Strikes Back? The Impact of Imperialism on Britain from the Mid-nineteenth Century* (Harlow: Pearson, 2005).

Tickner, Lisa, *The Spectacle of Women: Imagery of the Suffrage Campaign 1907–1914* (London: Chatto and Windus, 1987).

Till, Karen, *The New Berlin: Memory, Politics, Place* (Minneapolis: University of Minnesota Press, 2005).

Walkowitz, Judith, 'Going Public: Shopping, Street Harassment, and Streetwalking in Late Victorian London' *Representations* 62 (1998), 1–30.

Waller, Maureen, *London 1945: Life in the Debris of War* (New York: St Martin's Press, 2004).

Ward, Paul, '"Women of Britain Say Go": Women's Patriotism in the First World War', *Twentieth Century British History* 12/1 (2001), 23–45.

Ward, Paul, *Britishness since 1870* (New York: Routledge, 2004).

Warner, Michael 'Publics and Counterpublics' *Public Culture* 14/1 (2002), 49–90.

Webster, Wendy, *Englishness and Empire 1939–1965* (Oxford: OUP, 2005).

Weight, Richard, *Patriots: National Identity in Britain 1940–2000* (London: Macmillan, 2002).

White, Jerry, *London in the 20th Century: A City and Its People* (London: Vintage, 2001).

White, Jerry, *London in the 19th Century: A Human Awful Wonder of God* (London: Vintage, 2007).

Whybrow, Nicolas, *Art and the City* (London: I.B. Tauris, 2011).

Williams, Richard, *The Anxious City* (London: Routledge, 2004).

Winder, Robert, *Bloody Foreigners: The History of Immigration to Britain* (London: Abacus, 2004).

Index

BRITISH IDENTITIES SINCE 1707

The historiography of British identities has flourished since the mid-1970s, spurred on by increasing national consciousness in England, Scotland, Wales and Northern Ireland, and since 1997 by devolution. Historians and other academics have become increasingly aware that identities in the British Isles have been fluid and that interactions between the different parts of the British Isles have been central to historical developments since, and indeed before, the Act of Union between England and Scotland in 1707.

This series seeks to encourage exploration of identities of place in the British Isles since the early eighteenth century, including intersections between competing and complementary identities such as region and nation. The series also advances discussion of other identities such as class, gender, religion, politics, ethnicity and culture when these are geographically located and positioned. While the series is historical, it welcomes cross- and interdisciplinary approaches to the study of British identities.

'British Identities since 1707' examines the unity and diversity of the British Isles, developing consideration of the multiplicity of negotiations that have taken place in such a multinational and multi-ethnic group of islands. It will include discussions of nationalism(s), of Britishness, Englishness, Scottishness, Welshness and Irishness, as well as 'regional' identities including, for example, those associated with Cornwall, the Gàidhealtachd region in Scotland and Gaeltacht areas in Ireland. The series will encompass discussions of relations with continental Europe and the United States, with ethnic and immigrant identities and with other forms of identity associated with the British Isles as place. The editors are interested in publishing books relating to the wider British world, including current and former parts of the British Empire and the Commonwealth, and places such as Gibraltar and the Falkland Islands and the smaller islands of the British archipelago. 'British Identities since 1707' reinforces the consideration of history, culture and politics as richly diverse across and within the borders of the British Isles.

Proposals are invited for monographs and edited collections, including those that arise from relevant conferences.